Energy Transitions

Long-Term Perspectives

AAAS Selected Symposia Series

Published by Westview Press, Inc.
5500 Central Avenue, Boulder, Colorado

for the

American Association for the Advancement of Science
1776 Massachusetts Avenue, N.W., Washington, D.C.

Energy Transitions

Long-Term Perspectives

Edited by Lewis J. Perelman,
August W. Giebelhaus,
and Michael D. Yokell

AAAS Selected Symposium **48**

AAAS Selected Symposia Series

This book is based on a symposium which was held at the 1979 AAAS National Annual Meeting in Houston, Texas, January 3-8. The symposium was sponsored by the Society for the History of Technology and by AAAS Sections K (Social and Economic Sciences) and L (History and Philosophy of Science).

All rights reserved. No part of this publication may be reproduced or transmitted in any form or by any means, electronic or mechanical, including photocopy, recording, or any information storage and retrieval system, without permission in writing from the publisher.

Copyright © 1981 by the American Association for the Advancement of Science

Published in 1981 in the United States of America by
  Westview Press, Inc.
  5500 Central Avenue
  Boulder, Colorado 80301
  Frederick A. Praeger, Publisher

Library of Congress Cataloging in Publication Data
Main entry under title:
Energy transitions.
    (AAAS selected symposium ; 48)
    Includes bibliographies.
    1. Power resources. I. Perelman, Lewis J. II. Giebelhaus, August W. III. Yokell, Michael D., 1946- IV. Series: American Association for the Advancement of Science. AAAS selected symposium ; 48.
TJ163.2.E4937   333.79   80-21529
ISBN 0-89158-862-0

Printed and bound in the United States of America

About the Book

This book is among the first to examine the social, political, economic, and environmental dimensions of major long-term changes in the systems of energy supply and use. Providing a uniquely holistic perspective on the dynamics of energy and societal changes, the authors examine historical examples of major energy transitions--from petroleum and natural gas to either renewable resources or nuclear fission (or fusion); analyze the potential impact of the coming transition; and assess the implications of long-term energy transitions for present government energy policies.

About the Series

The *AAAS Selected Symposia Series* was begun in 1977 to provide a means for more permanently recording and more widely disseminating some of the valuable material which is discussed at the AAAS Annual National Meetings. The volumes in this *Series* are based on symposia held at the Meetings which address topics of current and continuing significance, both within and among the sciences, and in the areas in which science and technology impact on public policy. The *Series* format is designed to provide for rapid dissemination of information, so the papers are not typeset but are reproduced directly from the camera-copy submitted by the authors. The papers are organized and edited by the symposium arrangers who then become the editors of the various volumes. Most papers published in this *Series* are original contributions which have not been previously published, although in some cases additional papers from other sources have been added by an editor to provide a more comprehensive view of a particular topic. Symposia may be reports of new research or reviews of established work, particularly work of an interdisciplinary nature, since the AAAS Annual Meetings typically embrace the full range of the sciences and their societal implications.

> WILLIAM D. CAREY
> *Executive Officer*
> *American Association for*
> *the Advancement of Science*

Contents

Figures and Tables..................................*xi*

About the Editors and Authors.................. *xiii*

Introduction--*Lewis J. Perelman*......................1

PART 1. THE PAST

1   The Ascent of Oil: The Transition from
    Coal to Oil in Early Twentieth-Century
    America--*Joseph A. Pratt*........................9

    An Overview of Changing Patterns
       of Energy Use (1900-1920)   10
    Primary Markets for Fuel Oil   15
    The Advantages of Oil Over Coal   18
    Oil as an Alternative Energy
       Source: The Case of Texaco   23
    Conclusions: The Somewhat
       Useful Past   26
    References and Notes   29

2   Resistance to Long-Term Energy Transition:
    The Case of Power Alcohol in the 1930s--
    August W. Giebelhaus 35

    Historical Overview   36
    Alcohol and the Great Depression   37
    Oil Industry Opposition   39
    The Farm Chemurgic Movement   43
    The Atchison Experiment   45
    Final Attempts   52
    Conclusion   54
    References and Notes   56

Contents

3 An Attempt at Transition: The Bureau of Mines Synthetic Fuel Project at Louisiana, Missouri--*Arnold Krammer*........................65

    Early Synthetic Fuel Research   66
    The U.S. Bureau of Mines   69
    The Impact of World War II   70
    Post-War Developments   75
    The Louisiana, Missouri Project   77
    Mobilization of Opposition   90
    Epilogue   99
    References and Notes   100

4 The Transition to Solar Energy: An Historical Approach--*Ethan B. Kapstein*.........109

    Introduction   109
    Solar Pioneers   109
    Solar Water Heaters   113
    University Research   115
    Photovoltaics   118
    The Government Take-Over   119
    History and Policy   121
    References and Notes   121

PART 2. THE FUTURE

5 The Role of the Government in the Development of Solar Energy--*Michael D. Yokell*......127

    Introduction   127
    Should the Federal Government Subsidize Solar Energy?   127
    The Current Federal Solar Program   133
    Federal Subsidies for Solar Energy   134
    The Proper Balance Among Federal Programs to Subsidize Solar Energy   140
    Conclusion   142
    References and Notes   143

6 Forecasting Alternative Energy Futures: The Case of Solar Energy--*Gregory A. Daneke*....147

    Forecasting: Matters of Methodology   148
        A Typology of Philosophical Orientations, 149

	Energy Forecasting	153
	Solar Forecasts: Procedures in Search of Policies	155
	Toward a Teleological Approach	158
	The Non-Market Nature of Solar Energy,158; A Solar Policy Agenda, 162	
	Conclusions	165
	Notes and References	165
7	Patterns of Transition: Use and Supply of Energy--*Mark N. Christensen* 169	
	Elements of Consensus	170
	Implications for Energy Supply	174
	Implications for Energy Use	176
	Prospects for Future Demands for Energy	180
	Directions for Further Research	182
	Conclusions	182
	References and Notes	182
8	Speculations on the Transition to Sustainable Energy--*Lewis J. Perelman* 185	
	The Transition	185
	Feudalism and Theocracy	189
	The Psychology of the Transition	194
	Conflict	197
	Paradox	203
	The Madness of Schizophrenia	209
	Conclusions	211
	Notes and References	213

Figures and Tables

Chapter 1

Table 1	Specific energy sources as percentages of aggregate energy consumption, 1850-1955	10
Figure 1	"Handwriting on the Wall"	12
Table 2	Oil and coal as percentages of the total consumption of mineral fuels	13
Table 3	U.S. consumption of fuel oil by regions, 1909-1929	14
Table 4	U.S. consumption of fuel oil by uses, 1925 and 1928	16
Table 5	Growth of oil-burning vessels throughout the world: 1914, 1920 and 1922	20

Chapter 3

Figure 1	Distillation unit at the hydrogenation plant, Louisiana, Missouri	80
Figure 2	Process flow of coal hydrogenation plant	82
Figure 3	Flow sheet of Bureau of Mines modification of gas synthesis process	86
Table 1	Summary of unit-cost estimates of coal hydrogenation	94
Table 2	Comparison of estimated costs for coal-hydrogenation gasoline (1952)	96

xii   Figures and Tables

Chapter 5

Figure 1	Solar budget in relation to energy and federal budgets and to gross national product, FY78	132
Table 1	First- and second-best problems of solar energy and proposed solutions	136

Chapter 6

Table 1	A typology of forecast methodology	150
Table 2	Major solar energy forecasts	156
Table 3	Estimated cost of incentives used to stimulate energy production	160
Figure 1	Distribution of funding among various solar programs	161

About the Editors and Authors

Lewis J. Perelman *is a senior scientist in the Social Sciences Group at the Jet Propulsion Laboratory; his areas of specialization are solar energy, social impact assessment, program evaluation, and diffusion of innovations. A former program planner and policy analyst at the Solar Energy Research Institute, he has written extensively on resources, the environment, and social and behavioral change. He is the author of* Solar Energy Program Evaluation *(with D. Kline; Lexington, in press) and* The Global Mind: Beyond the Limits to Growth *(Mason Charter, 1976).*

August W. Giebelhaus, *an historian of economics, business, and technology, is an assistant professor at Georgia Institute of Technology. His primary area of research is energy history, and he is the author of* Business and Government in the Oil Industry: A Case of Sun Oil, 1876-1945 *(JAI Press, 1980) and general editor of* Dictionary of the History of the Petroleum and Petrochemical Industries *(Greenwood Press, forthcoming). Currently he is associate editor of* Technology and Culture.

Michael D. Yokell, *a specialist in energy and resource and environmental economics, is president and chief economist of Resource Management Consultants, Inc., in Boulder, Colorado. A former senior economist at the Solar Energy Research Institute and faculty member at the University of California-Berkeley and Washington State University, he is the author of several books, including* Yellowcake: The International Uranium Cartel *(with J. Taylor; Pergamon, 1979) and* Environmental Benefits and Costs of Solar Energy *(Lexington, forthcoming).*

Mark N. Christensen *is a professor in the Energy and Resources Program at the University of California-Berkeley and codirector of the multidisciplinary, DOE-funded study,*

"Distributed Energy Systems in California's Future." His publications include numerous articles and several texts in the field of geology.

Gregory A. Daneke is associate professor and director of the Environmental and Energy Planning and Administration Program at the University of Arizona. He has been a visiting professor in the Center for Technology Administration at The American University, a visiting scholar with the Energy Information Exchange Program of the U.S. State Department, and a member of the National Energy Planning Review Team of the U.S. General Accounting Office. He has published extensively on policy analysis, natural resources management, and energy systems assessment; most recently, Policy Analysis and the Public Interest (Allyn and Bacon, 1980) and Public Administration and Energy Policy (with G. Lagassa; D.C. Heath, 1980).

Ethan B. Kapstein is energy education director for the Massachusetts Audubon Society and a partner in the energy consulting firm, Butti, Kapstein and Perlin (Washington, D.C.; Santa Monica, CA). As a visiting scholar with the Department of Energy, he conducted research on the history, politics, and economics of solar energy and founded and edited the Energy History Report. In 1977-1978 he was a Fellow of the Rockefeller Foundation conducting research on solar energy in developing countries, and he currently holds a fellowship from the National Endowment for the Arts.

Arnold Krammer is a professor of history at Texas A & M University. A specialist in German industrial and foreign relations, he is principal investigator of the "German Document Retrieval Project," a study of Nazi Germany's industrial records on synthetic fuel and coal conversion for possible application to current U.S. energy problems. He has numerous publications in his field, including Nazi Prisoners of War in America (Stein & Day, 1979).

Joseph A. Pratt is an assistant professor of business administration at the University of California-Berkeley; his field of specialty is energy and business history. Currently he is a visiting scholar of the U.S. Department of Energy. He is the author of The Growth of a Refining Region: Twentieth Century Growth of the Houston Area (JAI Press, in press).

Energy Transitions
Long-Term Perspectives

Lewis J. Perelman

Introduction

Throughout most of human history, humankind has relied on the so-called "renewable" energy resources--wood, plants, animals, wind, and falling water--for its energy needs. In the nineteenth century, a major transition occurred to reliance on a fossil energy source: coal. The twentieth century has seen a second major energy transition from coal to petroleum and natural gas. Now another important energy system transition has begun. Within the next hundred years, the United States and other industrialized nations will have to make a new transition to reliance either on nuclear fission and fusion sources or, what now seems more likely, the traditional "renewable" resources.

The recent and continuing debate on national energy policies has focused almost exclusively on near-term problems, rather than on the requirements of the coming long-term energy transition. In the 1970s, energy policies have been as erratic and ephemeral as the transient crisis in fossil fuel supply and demand. Public and legislative interest in energy policy has fluctuated in inverse proportion to current inventories of gasoline, diesel fuel, heating oil, or natural gas. Transient supply constraints bring a flood of proposals for "crash" programs; subsequent supply surpluses bring a rapid relaxation to political apathy. On average, there has been relatively little social adjustment to the inevitable depletion of conventional fossil fuels. One reason for the lack of success in coming to grips with America's festering energy crisis may be the general lack of long-term vision of where the nation has come from and of where it could be going.

This book attempts to bring a longer-term perspective to the study of national energy policy than that of most of the existing, vast literature on the subject. The volume is far from being either flawless or comprehensive, but may be unique in its combination of disciplines and viewpoints. The authors represent the disciplines of history, economics, and the political, physical, and behavioral sciences; they simultaneously assess past, present, and future issues of national energy policy. Because the authors have collaborated only within the symposium at the 1979 AAAS Annual Meeting on which this volume is based, the degree of integration among the various essays is very limited. Nevertheless, this book at least should suggest the potential benefit of such multidisciplinary collaboration in long-term energy policy analyses. Certainly the authors have found even this limited collaboration a rewarding experience.

The focus of the book progresses more or less chronologically from the past to the near term to the more distant future. The contents are comprised of two parts. The four articles which comprise the first half of the book deal with historical cases relevant to energy transitions; they were solicited by August Giebelhaus on behalf of the Society for the History of Technology. The articles in the second half were solicited for the AAAS symposium by Michael Yokell and Lewis Perelman, and deal primarily with current and future energy policy issues.

Consistent with the overall concern with the coming transition from fossil fuels to renewable energy resources, six of the eight papers are concerned explicitly with solar energy (and conservation); the remaining two (by Pratt and Krammer) address topics in the history of fossil fuels with important relevance to future energy policy.

In "The Ascent of Oil", Joseph Pratt describes the first major American energy transition in the twentieth century: from coal to oil. Pratt's synopsis illuminates the historical roots of the petroleum industry's present dominance, and also suggests some of the difficulty in resurrecting coal as a replacement for oil and gas. Pratt observes that the success of oil in displacing coal as a principal energy source can be attributed largely to oil's economy and technical superiority in transportation and other applications. However, some additional reasons for oil's ascendance which are less commonly recognized are that oil was tied to the regional development of what we now call the "sun belt"; that oil's environmental superiority over coal was appreciated even in the 1920s; and that the petroleum industry already had developed an extensive production and marketing infrastructure from over 40 years of experience in the "illuminating oil" (kerosene) business before it began to compete with coal in the industrial market. Also, Pratt notes that the intense early competition both between coal and oil and within the oil industry led to inefficiencies in the use of both fuels which became a "legacy of waste" for future generations.

The two articles that follow, by August Giebelhaus and by Arnold Krammer, both provide early historical case studies which will leave the reader familiar with the current Washington frenzy over "synfuels" with an almost haunting sense of deja vu. Giebelhaus reviews "The Case of Power Alcohol in the 1930s," when the concept of converting agricultural produce into motor vehicle fuel was last in political vogue. Giebelhaus's detailed historical review shows why, despite substantial political support, the power alcohol movement of the 1930s was doomed from the start. The names, places, and principal actors may sound familiar; many are back in the news today with identical arguments both pro and con. In his conclusion though, Giebelhaus does note some of the charcteristics that make the contemporary "gasohol" movement different from the 1930s movement: limited domestic petroleum resources, automotive pollution regulations, and an increased emphasis on using agricultural waste rather than food crops as alcohol feedstock.

Krammer's article may be even more germane to recent proposals for massive federal investments to construct giant plants for the production of synthetic oil and gas from coal and oil shale. Krammer recounts the history of America's first coal-to-oil synthetic fuel demonstration plant, built and operated by the federal Bureau of Mines in the town of "Louisiana" in the state of Missouri from 1949 to 1953. The Louisiana project was undertaken in the belief that U.S. national security (in light of the experience of the Second World War) required development of a domestic alternative to imported oil, an alternative which could be provided best by synthesis of liquid fuel from the nation's abundant coal resources. At least according to the Bureau of Mines' estimates, the project demonstrated the feasibility of producing gasoline from coal at a cost (at that time) of less than 11¢ a gallon. Nevertheless, in 1953 the project was terminated for being "uneconomic." The lesson of this story is relevant not only to pending "synfuels" proposals, but to government solar energy programs, which also have been dismissed by some as "uneconomic" in the short term.

The last of the historical studies, by Ethan Kapstein, also serves as an introduction to the solar energy concerns of the book's second half. Kapstein's review of the history of solar energy development in the United States shows that important interest in the practical application of solar energy predates "Sun Day" 1978 by at least three-quarters of a century. Kapstein recalls research efforts on solar systems beginning at the turn of the century, and reminds us that a substantial solar water-heating industry existed in California and in Florida as early as 1904 and up until 1950. This historical experience in solar energy research and commercialization, Kapstein argues, deserves much more attention than it has received.

Shifting the focus to future-oriented energy policy, Michael Yokell's article on "The Role of the Government in the Development of Solar Energy" addresses the specific and critical issue of the use of government subsidies to accelerate the commercialization and diffusion of solar technologies. Yokell reviews the economic justifications for government intervention in the marketplace to advance the adoption of solar energy systems. Subsidies can respond to the problems of underpricing of conventional energy sources (which makes solar alternatives appear "uneconomic" in the short term), and of the risks involved in the development, production, and adoption of innovative technologies. Yokell evaluates a variety of specific subsidy options, and concludes by suggesting an optimum strategy for solar energy subsidization.

In "Forecasting Alternative Energy Futures," Gregory Daneke considers a second critical problem in solar energy policy analysis: the relationship between long-term forecasts and long-term planning. Daneke's wide-ranging review of energy forecasting efforts leads to the criticism that existing forecasting approaches underestimate and undermine the effectiveness of government policies to promote the development and adoption of solar technologies. The alternative approach Daneke recommends is "teleological forecasting": setting long-term goals according to social needs and values, then working backward to the present to define policy strategies capable of achieving the chosen goals.

Based on observations of recent social, political, and economic behavior related to energy, Mark Christensen attempts to induce the future "Patterns of Transition." Probing beneath the confusing energy controversies of current headlines, Christensen sees a growing societal consensus on three key points: first, prices of energy are increasing; second, environmental costs of energy production and consumption must be accounted for; and third, energy planning is plagued with uncertainties. As a result, the coming energy transition will be marked by growing constraints on expansion of centralized energy sources, and by an increased emphasis on conservation and decentralized energy sources. Christensen expects overall energy demand to level off and decline by the end of the century, and believes that end-users rather than producers will be the principal actors in the next energy transition.

In the final article, Lewis Perelman speculates about the potential consequences of the century-long future transition to a solar energy society. Perelman offers the provocative suggestion that the key characteristics of American society of the twenty-first century could be feudalism and theocracy. The author uses the term "feudalism" to refer to a specific set of social characteristics: wealth and power based largely on land holdings; political decentralization; a quasi-steady-state economy; and social stratification by caste and class. The feudalistic trend is seen as the result of the return to energy sources based on territory; and the theocratic tendency derives from the reduction in the secularizing effects of industrial growth based on fossil fuels. The principal concern of Perelman's analysis, though, is less with the outcome of the energy transition than with the intense social conflict expected in the transition process itself. Perelman concludes by calling for more investment in behavioral and social research that may reveal ways to reduce the destructive potential of the age of energy transition.

The editors are grateful to the other authors for their contributions to this volume. Our collaboration on the symposium and this resulting book was suggested originally by Authur Herschman of the AAAS. Elizabeth Zeutschel of the AAAS Meetings Office provided invaluable assistance in the planning of the symposium. The production services of the Solar Energy Research Institute Word Processing Center are much appreciated. We also appreciate the support of Joellen Fritsche of the AAAS Publications Office and of Frederick Praeger and Lynne Rienner of Westview Press in the development of this publication.

Part 1

The Past

Joseph A. Pratt

1. The Ascent of Oil: The Transition from Coal to Oil in Early Twentieth-Century America

History offers a useful, though often neglected, long-run perspective on current energy problems. Of particular interest to those who seek to analyze the coming transition from petroleum to alternative sources of energy is the history of the transition from coal to oil in the early part of the twentieth century. The first stage in the ascent of oil culminated after World War I, when the demand for gasoline expanded rapidly in response to a dramatic increase in the use of motor vehicles. But even before the surge in gasoline consumption in the 1920s, oil had begun to make inroads into several significant fuel markets traditionally served by coal. Indeed, the growing use of fuel oil in the first decades of the twentieth century was an important transitional phase in the decline of "King Coal" and the rise of its successor.

An examination of this expanding market for fuel oil from 1900 to 1930 reveals how an alternative energy source at the turn of the century--petroleum--survived and then prospered in competition with the predominant fuel of that era--coal. The study of this competition yields insights into both the historical strengths of the oil industry and the weaknesses of the coal industry which should suggest caution to those who currently underestimate the adaptability of the former while overestimating the potential dynamism of the latter. The comparison of the last energy transition to the present one reveals, however, more differences than similarities. The assumption of energy abundance in the earlier transition shaped political and economic attitudes that are strikingly different from those generated by current assumptions of energy shortages.

10 Joseph A. Pratt

An Overview of Changing Patterns of Energy Use (1900-1920)

The transition from coal to oil began only decades after coal had replaced wood as the primary source of energy in the United States. In the period of rapid industrialization after the Civil War, coal had established itself as the fuel of choice for the growing railroad system, for the most modern ships, for home heating in many cities, for the steel industry, and for numerous other manufacturing concerns. Aggregate statistics on national energy use suggest the rapidity and thoroughness of this shift to coal. In 1850 coal supplied less than 10 percent of the total energy consumed in

Table 1. Specific Energy Sources as Percentages of Aggregate Energy Consumption, Five-Year Intervals, 1850-1955.

Year	Estimated Total Energy Consumption (trillion BTUs)	Coal	Oil	Natural Gas	Fuel Wood
1850	2,357	9.3%			90.7%
1855	2,810	15.0			85.0
1860	3,162	16.4	.1%	n.a.	83.5
1865	3,409	18.5	.3	n.a.	81.2
1870	3,952	26.5	.3	n.a.	73.2
1875	4,323	33.3	.3	n.a.	66.4
1880	5,001	41.1	1.9	n.a.	57.0
1885	5,645	50.3	.7	1.5%	47.5
1890	7,012	57.9	2.2	3.7	35.9
1895	7,661	64.6	2.2	1.9	30.1
1900	9,587	71.4	2.4	2.6	21.0
1905	13,212	75.7	4.6	2.8	13.9
1910	16,565	76.8	6.1	3.3	10.7
1915	17,704	74.8	7.9	3.8	9.5
1920	21,378	72.5	12.3	3.8	7.5
1925	22,411	65.6	18.5	5.3	6.8
1930	23,705	57.5	23.8	8.1	6.1
1935	20,456	52.0	26.9	9.4	6.8
1940	25,235	49.7	29.6	10.6	5.4
1945	32,700	48.8	29.4	11.8	3.9
1950	35,136	36.8	36.2	17.0	3.3
1955	40,796	28.7	40.0	22.1	2.6

Source: Sam Schurr, Bruce Netschert, Energy in the American Economy, 1850-1975 (Baltimore: Resources for the Future, Inc., 1960), pp. 36, 145. Reprinted by permission.
n.a.: Not available.

the U.S.; by 1900 this figure had jumped to more than 70 percent (see Table 1). During these years, coal became firmly established in the fuel markets of the nation's largest urban-industrial centers, which were then located almost exclusively in the Northeast and the Midwest and thus enjoyed easy access to the major sources of coal in the Appalachian field. By 1900 coal accounted for more than 93 percent of all mineral fuels consumed in the U.S., and abundant supplies of known reserves seemed to assure that King Coal would have a long reign as the dominant source of energy in a national economy characterized by a rapidly increasing demand for energy [1].

Yet even as the coal industry looked forward to a new century of dominance, events in the oil industry foreshadowed its future decline. In 1900 and 1901, the discovery of vast new oil fields in Texas and California greatly expanded the marketable supply of crude oil. These new sources of crude, however, were not well-suited for the production of "illuminating oil" (kerosene) which was then the primary product derived from petroleum. In order to sell the oil from the new fields, several young companies--most notably Texaco and Gulf Oil in Texas and Standard of California and Union Oil in California--aggressively pushed the use of oil as fuel [2]. As early as 1902, some of the more enthusiastic proponents of fuel oil--including those who sought to attract investments in oil companies--were already forecasting the ultimate triumph of oil over coal as the dominant source of fuel (see Figure 1). Such forecasts initially elicited only ridicule from spokesmen for the coal industry, but the steady expansion of fuel oil sales gradually made petroleum an important energy alternative. Indeed, the estimated consumption of refined fuel oils and crude oil used as fuel expanded from 6.4 million barrels in 1899, to more than 91 million barrels in 1909, and to more than 300 million barrels in 1920 [3]. This impressive record of growth allowed oil to register significant gains relative to coal (Table 2). Although coal remained the dominant source of energy in 1920, fuel oil had already become a major new source of energy. During the next three decades the steady expansion in the use of fuel oil, gasoline, and natural gas accelerated the decline of coal and the ascent of oil [4].

Although coal's relative share of the total consumption of mineral fuels declined significantly between 1900 and 1920, this did not necessarily mean that coal was steadily losing markets to its rival. In fact, the consumption of bituminous coal more than doubled in this period. So great was the demand for all forms of energy that the markets for coal and oil both expanded dramatically, especially in the first decade of the century, a period that witnessed the most

Figure 1. "Handwriting on the Wall." <u>National Oil Reporter</u>, Volume 3, #10 (October 16, 1902), p. 13.

Table 2. Oil and Coal as Percentages of the Total Consumption of Mineral Fuels (Based on BTU Value)

	Oil (percent)	Coal (percent)
1900	3.1	93.4
1905	5.6	91.1
1910	7.1	89.2
1915	9.2	86.4
1920	13.9	81.6

Source: Sam Schurr and Bruce Netschert, Energy in the American Economy, 1850-1975. (Resources for the Future, Inc., Baltimore, 1960), pp. 514-515.

rapid increase in energy use of any decade in the modern history of the United States [5]. As total energy consumption more than doubled between 1900 and 1920, the booming oil industry had ample opportunity to expand without directly challenging King Coal in many of its traditional strongholds in the northeastern and midwestern industrial centers.

Even national statistics that show oil's steadily increasing proportion of the total energy supply do not capture the total significance of the rise of oil in the first two decades of the century, for the initial ascent of oil was largely a regional, not a national, process. The transition to oil came most thoroughly and most quickly in the major population centers of the West and Southwest. These "Sunbelt" regions were growing rapidly even before the discovery of vast new sources of oil in California, Texas, Oklahoma, and Louisiana, but oil accelerated their industrial growth by providing an inexpensive source of fuel for both manufacturing and transportation. In part because of the impetus provided by this new source of fuel, the Far West and the Southwest grew far more quickly than any other sections of the nation in the two decades after 1900, and provided an ever-expanding market for the oil and natural gas which fueled their development [6]. The fate of these new fuels and that of the emerging Sunbelt were thus tightly intertwined at an early and crucial juncture in the development of both.

Much of the increased consumption of fuel oil did not, therefore, result from the direct substitution of oil for coal. Instead, oil took advantage of the opportunities afforded by the dramatic expansion in energy use throughout the nation and by the especially rapid industrialization of the West and Southwest, where development had previously been

Table 3. United States Consumption of Fuel Oil by Regions, 1909-1929. (Million U. S. Barrels)

District	1909	1926	1927	1928	1929
Atlantic Coast region					
New England	—	20.9	20.9	18.7	21.8
Central District[a]	--	77.1	77.1	82.3	90.5
Southern District[b]	—	12.3	9.9	7.9	8.2
Total	12	112.3	110.1	111.8	125.3
South Central region					
Eastern[c]	—	1.9	1.2	1.5	--
Western[d]	--	89.5	83.8	85.9	—
Total	22	93.4	87.0	89.4	89.0
North Central region					
Eastern[e]	—	19.0	24.4	28.6	--
Western[f]	--	2.5	2.9	3.1	—
Total	8	25.5	31.3	36.7	40.0
Rocky Mountain region	--	6.8	7.2	6.0	6.0
Pacific Coast region					
California, Arizona, Nevada	--	90.9	93.3	96.5	94.8
Oregon, Washington	--	13.5	13.5	17.6	18.1
Total	34	104.4	106.8	114.1	112.9
Grand total	76	342	342	358	373

Source: National Industrial Conference Board, Oil Conservation and Fuel Oil Supply (New York: National Industrial Conference Board, Inc., 1930), pp. 77, 160.

[a] From New York to Virginia inclusive.

[b] From North Carolina to Florida inclusive.

[c] Alabama, Mississippi, Tennessee.

[d] Texas, Louisiana, Arkansas, Oklahoma, Kansas, Missouri.

[e] Kentucky, Ohio, Indiana, Illinois, Michigan, Wisconsin.

[f] Minnesota, Iowa, Nebraska, North and South Dakota.

The Ascent of Oil   15

constrained by the lack of access to coal. But the use of fuel oil did not remain isolated in these Sunbelt regions, for the advantages of oil over coal for a variety of tasks gradually became clear as the discovery of new oil fields continued to expand the amount of petroleum available for fuel.

Primary Markets for Fuel Oil

Statistics on fuel oil consumption are sketchy at best for the period before World War I, but a systematic survey in the 1920s by the U.S. Bureau of Mines provides a starting point for an examination of the primary uses of fuel oil throughout the first decades of the century. Table 3 shows the extent of fuel oil use by oil companies, shipping, railroads, and public utilities in the mid-1920s; these industries constituted the major markets for fuel oil before the 1920s.

Railroads consumed a substantial percentage of all fuel oil sold throughout this period. In the West and Southwest the conversion of railroads to oil began almost immediately after the discovery of the large new Texas and California fields [7]. Simple economics dictated this shift. Trains required large amounts of fuels, and the sudden availability of a much less expensibe alternative to coal presented an excellent opportunity for lowering operating expenses. The cost of conversion was low, especially in light of the immediate and substantial fuel savings achievable with fuel oil. Unlike that for other potential consumers, the cost of oil to the major railroads was not greatly affected by the ebb and flow of production in one particular field, since railroad tracks extended throughout the West and Southwest, assuring the companies of relatively easy access to any newly discovered oil fields. In addition, ownership of oil-producing lands and substantial investments by some railroad owners in oil companies gave many of the major railroads in the Sunbelt strong operating ties to that region's burgeoning young oil industry.

These railroads generally converted to fuel oil very quickly. From 1900 to 1905, for example, the fuel oil consumption of the Southern Pacific Railroad's operations west of El Paso rose from about 100,000 barrels annually to more than 5 million, while its consumption of coal dropped to about half of its 1900 total of more than 1.2 million tons [8]. This change was permanent, and it occurred on railroads throughout the Sunbelt. By 1926 railroads were the major market for fuel oil in the West and the Southwest. In older industrial sections of the nation, oil did not displace coal as the primary fuel for railroads until the rapid switch to

Table 4. United States Consumption of Fuel Oil by Uses, Including Crude, Gas and Fuel Oils, 1925 and 1928. (In million barrels of 42 U. S. gallons-- figures to the nearest million).

	1925	1928
Used on oil leases	4	4
Used by pipe leases	2	2
Used as refinery fuel	51	46
(a) Total used by oil companies	57	52
Used by United States Navy	7	7
Used for domestic bunkers	30	36
Used for foreign bunkers	43	51
(b) Total marine use	80	94
Used by locomotives	60	63
Used by railroad shops and ferries	11	8
(c) Total railroad use	71	71
Used by electric power companies	10	7
Used by gas companies	23	23
(d) Total public utilities	33	30
Used for commercial heating	12	17
Used for domestic heating, including distillate	9	14
(e) Total heating use	21	31
Used by iron and steel industry	--	19
Used by mines and smelters	--	7
Used by cement and ceramic industries	--	8
Used by food industries	--	6
Used by paper, pulp, and lumbering industries	--	5
Used by textile industries	--	5
Used by all other industries	--	29
(f) Total other manufacturing and industrial uses	77	79
Total all uses as above	339	357

Source: National Industrial Conference Board, Oil Conservation and Fuel Oil Supply (New York: National Industrial Conference Board, Inc., 1930), p. 159.

The Ascent of Oil 17

diesel fuel after World War II. Of course, in the intervening years, oil's share of transportation-related energy consumption rose dramatically throughout the nation as motor vehicles made rapid inroads into the railroad's traditional dominance of transportation [9].

Another form of transportation, steamships, provided a second major market for fuel oil. Tankers for carrying petroleum products were the first large oceangoing vessels to make the switch from coal-burners to oil-burners, and in the first decades of the century, they were followed by other vessels sailing from the West and Gulf Coasts [10]. The construction of oil bunkers around the world to service these ships gradually removed one final uncertainty that had caused some shipowners to hesitate earlier. In the years before World War I, the U.S. Navy gradually made the transition from coal to oil, and this highly publicized choice undoubtedly encouraged other shippers to consider oil-burners [11]. The Navy's decision signaled the coming of age of fuel oil in shipping, since it proclaimed the superiority of oil in the most critical of all markets. The Navy's judgment was vindicated by the performance of its ships in World War I, and after the war the shift to oil-burners greatly accelerated (see Table 4). As with the railroads, the price of oil was a prime reason for the shift, but convenience, superior performance, ease of handling and savings in shipping space were also strong incentives to convert existing steamships from coal to oil or to construct new ships with oil burners.

Fuel for tankers was only one of several uses that oil companies made of their own products. Far more important was the use of fuel oil to provide the intense heat needed for petroleum refining. Their own substantial energy needs thus gave the vertically integrated oil companies one large, expanding market, and an estimated 5-10 percent of all oil produced in this period was burned in the refineries [12]. For awhile, some major refiners simply burned crude oil under their stills, but gradually refined fuel oils replaced crude. In the mid-1920s, the rising value of gasoline and the greater availability of natural gas encouraged many refiners to switch from fuel oil to natural gas, and many other industrial users gradually followed the refiners' lead in converting to this new fuel [13].

Table 3 above provides a summary of fuel oil use in the 1920s in the most important of these industrial markets. In several major industries (including iron and steel, cement, textiles, food, and paper and pulp), in home heating, and in electric utilities, fuel oil gradually captured a share of markets that had previously been supplied almost exclusively by coal. Then beginning in the 1920s, oil gradually relinquished some of these markets to natural gas, which was more a "companion" fuel than a competing fuel, since it was bound to the oil industry by strong geological, financial, and institutional ties. The growing use of natural gas sustained the transition away from coal at the same time that it displaced fuel oil from some markets, thereby freeing petroleum to meet the rising demand for gasoline [14].

In the West and Southwest, oil attracted many more industrial users than in other parts of the nation. Indeed, soon after the discovery of the large California and Texas oil fields at the turn of the century, fuel oil had captured most industrial markets in the major cities of Texas and California [15]. More gradually, it made inroads into some such markets on the East Coast as well as in the home heating markets there. Most of these early markets for fuel oil continued to expand after the 1920s, and in combination with the accelerating consumption of gasoline, they pushed oil and natural gas past coal as a source of energy by 1950.

The Advantages of Oil Over Coal

The rapid ascent of oil in the early decades of the century reflected both the strengths of oil and the weaknesses of coal. A comparison of the economic and institutional factors influencing the relative development of the two industries suggests why the use of oil grew so rapidly in this early period and throughout the rest of the century.

The crucial determinants of the ascent of oil were economic [16]. In an era of energy abundance, the relative prices of oil and coal were the primary determinant of the choice of fuel for consumers, and the sum of these choices added up to a national "energy policy." When and where oil could lower the cost of energy to a particular consumer, it was generally substituted for coal. The price at which each fuel could be delivered to particular classes of customers in each section of the country thus determined the broad outlines of the last energy transition.

In the years immediately after the discovery of oil in Texas and California, an average price of 25¢ to 30¢ per barrel made oil a spectacular energy bargain for those fortunate enough to have access to the new fields [17]. The Southwest, the West, and those parts of the East Coast that could be served by the existing tanker fleet enjoyed a two- or three-year period in which the price of oil was so low in comparison to coal that many were willing to gamble on this as yet unproven new source of fuel. The costs of converting coal-burning equipment to oil was generally quite low, as was the investment in storage facilities for oil [18]. Some large fuel users could therefore pay the entire expense of conversion and still pocket substantial fuel savings in their very first year of oil use. This provided a strong economic incentive for fuel-intensive industries to change quickly to oil. Even if oil gradually became much more expensive, the short-run profits made from burning cheap oil would usually more than cover the costs of converting from coal to oil and even back to coal.

Few who tried oil, however, subsequently returned to coal. Even when the price of oil rose to between 70¢ and $1.10 per barrel at the well in the years before World War I, its low transportation costs to the major fuel markets in the nation allowed oil to remain price competitive with bituminous coal, which sold in the range of $1.05 to $1.30 per ton at the mine throughout these years [19]. Since most consumers figured that three and one-half barrels of oil equaled about one ton of coal for heating purposes, oil competed effectively only when and where its transportation costs were substantially less than those of coal. But oil enjoyed an important technological advantage, since it could be transported in tankers and pipelines much more cheaply than coal could be shipped by rail. Indeed, transportation costs often doubled the price of coal delivered from the Appalachian fields to the eastern seaboard, and the delivered price of coal to the West or Southwest could amount to triple or even quadruple the cost of coal at the mine. In these key regions, the cost of transportation thus became the prime competitive advantage of oil, and the growing oil companies strengthened their position vis-a-vis coal by investing heavily in pipelines and tankers [20].

Since transportation costs made coal prohibitively expensive in the West and Southwest, these oil-rich and coal-poor sections of the country led the way in the transition to fuel oil. In particular, the growing urban-industrial centers of Los Angeles, San Francisco, Houston, and Dallas became almost totally dependent on oil to supply their expanding energy needs. Since the combined populations of these four cities grew from slightly more than one-half million in 1900 to almost 1.4 million by 1920, they presented a substantial and growing market for fuel oil [21]. Neither Houston nor Los Angeles had developed much of an industrial base before the discovery of oil in Texas and California, so these two cities--which became the fastest growing areas in the nation for most of the twentieth century--especially industrialized around petroleum. On the other hand, in San Francisco, fuel oil challenged coal for existing markets as well as supplying many new markets, and coal could neither

Table 5. The Growth of Oil-Burning Vessels Throughout the World: 1914, 1920 and 1922

Nationality	1914 No.	1914 Gross Tons	1920 No.	1920 Gross Tons	1922 No.	1922 Gross Tons
American	239	656,364	1326	6,059,273	1790	8,857,087
British	121	558,743	338	1,822,444	601	3,460,428
Dutch	42	90,632	92	250,460	153	592,578
French	4	15,388	21	73,836	57	245,761
Italian	3	2,092	33	106,471	56	192,256
Norwegian	7	33,963	97	338,737	175	668,819
All others	85	366,575	117	388,028	278	987,619
Total	501	1,721,747	2021	9,039,247	3110	15,004,543

Exclusive of Army, Navy, Admiralty, and other government oil burners; but including oil burners on Great Lakes.)

Source: U.S. Bureau of Mines, Pollution by Oil of the Coast Waters of the United States (Washington, 1923), Appendix C.

capture these new markets nor defend its traditional ones. In 1900, the city's coal consumption reached almost two million tons, much of which it imported from Canada and Australia. Despite its rapid population growth after 1900, by 1914 San Francisco consumed only 400,000 tons of coal. Throughout the area from the Gulf Coast of Texas to the Pacific Coast a similar process occurred, with oil displacing coal from traditional or potential markets [22].

In locations such as Seattle, New Orleans, and the cities on the eastern seaboard, oil faced much stiffer competition, since these cities were nearer to the major supplies of coal and farther from the large new oil fields. New Orleans, the largest city in the South at the turn of the century, witnessed especially intense competition between the two fuels, as Pittsburgh and Alabama coal shipped down the Mississippi River competed with coastal shipments of Texas oil. The result was something of a standoff; oil captured many new markets while coal retained substantial markets and even recaptured customers who had switched to oil in the first flush of conversion after the discovery of the Texas oil fields [23]. As on the East Coast, however, the competition of coal and oil in New Orleans revealed that oil's advantages were more than strictly economic. Many of those who switched to cheap oil were later willing to pay a premium for oil rather than to switch back to coal. These consumers cited the quality of the heat produced, the ease of handling oil, and the lack of offensive smoke as justification for their continued reliance on oil. The efficiency and convenience of oil assured it of a growing market in those areas where it could stay roughly price competitive with coal.

The location of the major coal and oil fields and the cost of transportation thus brought the two fuels into direct and vigorous competition in several regions, while at the same time insulating oil from competition with coal in several of the most rapidly industrializing sections of the country. Table 5 shows the results as of the mid-1920s. California and Texas remained the most important customers for fuel oil, with substantial consumption also recorded along the much more populous Atlantic Coast.

Economics--as conditioned by geography--thus largely determined the pace and timing of the transition from coal to oil. Even when and where oil enjoyed a price advantage over coal, however, many potential consumers hesitated to shift to oil because of a commonly voiced concern over the uncertainty of future supplies of oil. Prominent spokesmen from the oil industry, from government, and from universities periodically fed this fear with pronouncements of an impending shortage of oil [24]. But every new cry of "wolf" seemed to bring with it the discovery of vast new sources of oil, and consumers learned to disregard the warnings of experts and heed instead what seemed to be a lesson of experience: neither petroleum products nor prophecies of oil shortages were ever likely to be in short supply for any length of time. Ironically, despite the vast supply of known coal reserves, shortages of this fuel were more likely to directly affect consumers. From 1902 through the mid-1920s, a series of violent strikes periodically disrupted the supply of coal. The oil industry, which was much less labor-intensive than coal, seldom experienced such prolonged or effective strikes, and the short-run uncertainties caused by coal's recurring labor problems were more pressing to both industrial and domestic consumers of energy than were speculations about the long-run sufficiency of the supply of oil [25].

In comparing the prospects of the two industries, a potential consumer could hardly fail to note the differences in their structure and performance. In comparison to the companies that sold bituminous coal, the major oil companies were paragons of efficiency, dependability, and innovation. They grew quickly by competing aggressively, both with the coal companies and with each other, and for the consumer this generally meant good service at decreasing prices. These companies were confident of their own futures and that of their industry, and they were thus willing to offer fuel oil at or even below cost in order to induce a consumer to try their product, one which they felt was superior to coal as a source of energy. The major oil companies were generally well-financed and well-managed, and they quickly grew into vertically integrated, internationally active firms that had the resources to improve their product and their service. This was in sharp contrast to conditions in the "sick" bituminous coal industry, which Herbert Hoover properly labeled "the worst functioning industry in the country" in the 1920s. The small, financially weak, labor-intensive companies which made up the bituminous coal industry seldom had the resources to develop innovations in either production or marketing, and they were hard-pressed to compete with the expansive petroleum industry [26].

Despite its weaknesses, the coal industry might well have continued to expand had it faced a competitor that sought to organize an alternative energy industry from the ground up. Instead, a thriving industry based on the production and sale of kerosene for lighting diversified into a closely related market for fuel oil. This "new" energy industry thus inherited proven techniques for producing, transporting, refining, and marketing its products, an efficient organizational model of vertical integration, and investment capital generated by the previous success of the oil industry. The new companies that pushed the use of fuel oil in the Southwest and the West generally adapted existing administrative and technical practices with relatively minor adjustments, and this enabled them to expand rapidly even in their formative years.

At times the growing young companies in California and Texas even made direct use of the facilities of their older, more developed rival, Standard Oil (New Jersey). Standard's requisition of one of the major California companies, the Pacific Coast Oil Company, represented a formal combination of resources and experience gained largely in the illuminating oil industry with opportunities and new markets generated by the rising West Coast fuel oil industry. Strict antitrust laws in Texas prevented such direct involvement there by Standard. Yet as the following account of the early history of Texaco will illustrate, Standard nonetheless fostered the development of the Texas oil industry in a variety of important ways [27].

Building on a foundation laid by over forty years of experience in producing oil for illumination, the new oil companies, as well as Standard Oil, quickly constructed an industry capable of supplying an alternative source of energy that could successfully compete with coal. The technical and organizational capacities subsequently developed in the production of fuel oil enabled these same companies to make an equally rapid adjustment several decades later when the rise of the automobile caused a basic shift in the national pattern of energy use.

Oil as an Alternative Energy Source:
The Case of Texaco

The transformation of the petroleum industry from the production of kerosene to the production of fuel oil, and then to gasoline was at the heart of the transition from coal to oil. By adapting an existing industry to changing conditions, dynamic young companies in California and Texas led the way in developing new markets for oil as energy. A closer look at the history of one of these companies, Texaco, suggests how a relatively small company, producing an as yet unproven alternative energy source, grew to become a major company in the first two decades of its existence [28].

Texaco grew out of Spindletop, which was the first giant Texas oil field. The discovery of oil at Spindletop in 1901 set off a scramble for wealth which attracted oil men and would-be oil men from throughout the nation. Amid the resulting chaos, the predecessor company of Texaco was chartered in 1902 with an authorized capitalization of only $50,000. But this relatively small firm had two important attributes that allowed it to attract the attention of large investors: experienced, respected management and substantial proven oil reserves. By 1903 the company had been reorganized with authorized capital of $3 million. With these additional investment funds, Texaco searched for still other sources of crude oil while gradually building the transportation, refining, and marketing capacities required to become a major producer of fuel oils. Good luck, opportunistic management, and the ability to use crude oil reserves to attract investment all helped Texaco to expand rapidly in its formative years.

Texaco's early success also reflected the previous development of the oil industry and especially of Standard Oil, which had exercised near monopoly control over all branches of the petroleum industry at the turn of the century. Before the discovery of the new flush fields in Texas and California, Standard had built an efficient, tightly organized, vertically integrated company devoted largely to the production of kerosene from crude oil. In its efforts to market oil for use as energy rather than as illumination, Texaco copied much of the technology employed by Standard while gradually developing a vertically integrated organizational structure patterned after that created by John D. Rockefeller in the late nineteenth century.

In attempting to replicate much of Standard's operation, Texaco made good use of numerous direct ties to the established company. In its formative years, Texaco relied heavily on former employees of Standard Oil. Its first president as well as many of its early managers and skilled workers had learned their trades as employees of the giant firm. In addition, much of the working capital needed by Texaco to develop new markets came from the sale of large shipments of crude oil to Standard, which made use of its expanding flet of tankers to transport this Texas crude to urban-industrial centers on the East Coast. As Texaco moved into refining, its marketing specialist even solicited and received an analysis of the quality of its earliest refinery runs from experts at Standard Oil, which in return received the option to purchase the products from subsequent refinery runs before Texaco offered them to other buyers. Clearly, Texaco's quick emergence as a major supplier of fuel oil owed much to its ability to draw on the experience of an established company [29].

Texaco's early history thus reinforces a central point: as fuel oil began to challenge coal, it was not an entirely new industry. The primary problem facing those companies that moved into the fuel oil business was not how to fashion completely new organizations while perfecting new technologies of production. Instead, these companies faced the much less formidable tasks of modifying existing refining processes and of finding new markets for their product. Neither problem proved particularly difficult. The heavy crude produced in Texas and California often required little or no processing before being sold as fuel oil, and the major regional markets surrounding the oil fields in these states proved most eager to adopt this inexpensive, efficient new source of energy.

In competing for these markets, Texaco displayed a competitive zeal that hastened both its own growth and the general adoption of fuel oil. One example will serve to illustrate the competitive climate in these early years of the confrontation between coal and oil. The large sugar refineries along the Mississippi River in Louisiana were major purchasers of coal at the turn of the century, and Texaco went to great lengths to break into this market. It first hired a longtime coal merchant from New Orleans to guide these efforts, securing the relationship by making available a block of its very valuable stock to the new employee. His detailed knowledge of the fuel situation of the major sugar refiners gave Texaco a competitive edge in bidding for new business. If necessary, the company initially offered its fuel oil to these refiners at little or no profit in order to promote this new source of energy. Texaco assumed correctly that after demonstrating the superior efficiency and convenience of fuel oil, it could later raise its prices. Although many refiners switched to fuel oil, Texaco's competitors garnered some of the resulting sales by undercutting the firm's prices [30].

Texaco and its rival oil companies showed a similar zeal in going after markets throughout the Southwest. These companies were as aggressive in their competition with each other as they were in their efforts to capture existing markets from the coal industry. In this early era of fuel oil sales, a glut of crude oil encouraged intense competition among the numerous new oil companies trying to survive in the southwestern fields. Further heightening the competition to sell fuel oil was the fact that much of the crude oil from these fields could not be easily refined into kerosene or gasoline with existing technology. The result—at least in the short run—was very low prices for fuel oil in the Southwest. Given the high cost of transporting their product, coal companies in Pennsylvania and Alabama simply could not meet these prices.

Texaco and the other large firm that emerged from the Texas fields, Gulf Oil, had an even stronger incentive to cut costs and lower prices. They sought to break into East Coast markets in competition with Standard Oil. The low cost of tanker transportation from the Gulf Coast to the eastern seaboard made such competition feasible, and the aggressive marketing and pricing policies of Texaco and Gulf insured success. Standard Oil was no stranger to cutthroat competition, and it defended its traditional territory by vigorously developing its own fuel oil distribution network. Thus, more than any of the individual firms, fuel oil itself emerged as the victor in this struggle, as low prices and more frequent deliveries won many new converts from coal to oil [31].

These companies did not, of course, go away empty-handed. Fuel oil sales in the Southwest and on the East Coast helped Texaco and Gulf Oil expand with heartbreaking swiftness in their formative years, thus placing them in a good position from which to move into the production and sale of gasoline in the years immediately before and after World War I. In 1917--only sixteen years after their founding--Texaco and Gulf ranked, respectively, as the twenty-fourth and twenty-sixth largest industrial concerns in the nation. Both companies counted total assets of almost $150 million, a figure that ranked them ahead of the largest coal company then in operation [32]. The ability of these and other western and southwestern oil companies to develop new markets for fuel oil and to challenge coal for existing energy markets was the crucial determinant of the pace of the early transition from coal to fuel oil. The Texaco story suggests how aggressively and how successfully these growing companies pushed the transition.

Conclusions: The Somewhat Useful Past

History does not teach lessons that offer solutions to current energy-related problems, but it does provide insights that are useful in understanding an on-going process of energy-related change. The study of the last energy transition suggests several conclusions that help place current concerns in a broader context.

Of particular interest today are historically based comparisons of the performance of the oil and coal industries. In the early decades of the century, oil clearly outperformed its rival. Even before its entry into the energy field, the oil industry had developed important administrative and technical strengths that its once and perhaps future rival could not match. The oil industry's special advantages were its efficient transportation system and its convenient, easy-to-use fuel. Although oil gradually became established throughout the national economy, it also continued to possess an important geographical advantage over coal, since it enjoyed nearly total dominance in the rapidly industrializing Sunbelt. Even in the earliest years the most discussed weakness of oil was the uncertainty of future supplies, but then--as throughout most of the century--the discovery of the new oil fields generally allayed this concern. As early as the second decade of the century, large amounts of imported oil from Mexico filled some of the growing demand for fuel oil, and even after troubles in Mexico clearly illustrated the uncertainties that accompanied reliance on foreign sources of crude oil, American consumers accepted this additional uncertainty rather than reassessing their commitment to oil [33].

In addition to the strengths of the petroleum industry, the ascent of oil reflected the weaknesses of coal. In contrast to oil, an assured long-run supply was perhaps the only thing that was certain about the operation of the coal industry. Developed supplies, however, were largely concentrated in the older industrial regions of the East, and high transportation costs blocked competition with oil for many of the new markets that emerged in the twentieth century. Financially and administratively, coal remained disorganized, and it displayed little evidence of technical creativity. An additional problem that initially seemed minor has grown in importance over time. Even in the early years of the ascent of oil, many of those who switched from coal cited oil's "cleanliness" or the "lack of bothersome smoke or dust" as a reason for the change. All in all, the competition between oil and coal pitted a healthy industry against a sick one, and the results were (and perhaps still are) predictable.

The competition between the two fuels was largely unrestrained, with the "free market" dictating the ultimate outcome. In this historical energy transition, the interplay of economic forces displayed the great strength of a price system by providing a decentralized decision-making mechanism that allowed for a very rapid transition of a new and less expensive source of energy. As it rewarded these companies, it also disciplined their behavior with competitive pressures that encouraged lower prices and higher quality.

Important weaknesses of the "free market" as a guide to the transition to a new energy source were also apparent to observers at the time as well as to historians. In particular, a great deal of waste resulted from the unrestrained competition between these two energy sources. John Ise, an economist who specialized in the study of natural resources, surveyed this history of waste in 1926 and concluded that:

> We, or our children, will some day, in a time of national peril, look back regretfully at the wanton waste of oil and oil products in these days of plenty, and we will perhaps squander vast sums of money in bootless efforts to secure satisfactory substitutes for material which we have wasted [34].

In the intensely competitive fuel markets of that time, expanding markets, not the efficient use of energy, was the primary concern of the companies. "Low" uses for both fuels resulted, including the sale of crude oil and unprocessed coal for fuel. This made short-run economic sense to the individual companies involved, but it proved costly to future generations of energy consumers.

This narrow definition of efficiency from the short-run perspective of the individual firm, which treated coal and oil as competing products rather than as two complementary sources of a scarce and dwindling supply of energy, was an important legacy of the last energy transition. No one with power in industry or government based decisions on a broader definition of efficiency that included a long-run perspective of the nation's energy needs.

One possible source of this missing long-run view was the government. Perhaps the most striking difference between the last energy transition and the present situation is the fact that the public sector took a relatively limited role in the early years of the shift from coal to oil. State railroad commissions in both Texas and California made important rate decisions that favored oil over coal, but beyond these decisions, government agencies exercised little influence. An assumption of economic and energy abundance and a political tradition of limited government combined to make the early shift from coal to oil largely a private matter. The government's lack of either the institutional capacity or the ideological rationale for monitoring--much less guiding--the last energy transition suggests the extent of the historical weakness that the public sector has been attempting to overcome in the 1970s. Government came to this complex problem late in the process of change, and it then faced the difficult and as yet unfinished tasks of creating new institutions and new sources of information with which to define the government's role amid the conflicting pressures generated by an open, democratic political system.

The study of the last energy transition also places the current attitudes and behavior of the large oil companies in historical perspective. These corporations developed what could best be described as the "habit of power" during their ascent to energy dominance. This power rested squarely on a history of sustained expansion of output in which consumers and voters rarely demanded more than inexpensive, abundant supplies of energy. Until the very recent past, the major oil companies have had very little experience in sharing power over decisions that affect their industry, and they have retained a strong suspicion of government involvement in what has traditionally been their private domain.

The ascent of oil thus took place in a much less complex environment than that now facing energy alternatives to oil. Abundant supplies of several forms of energy gave the nation the luxury of choosing an energy source primarily on the basis of its short-run economic cost to individual consumers. Competition within each energy industry and among the different industries assured the rapid development of energy resources--at least for as long as large domestic supplies of easily exploitable resources remained. The widespread waste that accompanied the last energy transition resulted from the unfettered exuberance of an adolescent industrial nation that gave little thought to the long-run sufficiency of energy supplies. The United States now faces another energy transition, but scarcity and not abundance is the underlying assumption that will shape the movement away from oil and toward other sources of energy. The habits ingrained by decades of excesses, however, have been most difficult to alter, and even ardent calls for a new austerity in the use of energy are often delivered with a tinge of nostalgia for the energy world we have lost.

References and Notes

1. For a detailed examination of trends in energy use in the past, see Sam H. Schurr and Bruce Netshert, Energy in the American Economy, 1850-1975: An Economic Study of Its History and Prospects (The John Hopkins Press, Baltimore, 1960). A good overview of these trends is found in Hans H. Landsberg and Sam H. Schurr, Energy in the United States: Sources, Uses and Policy Issues (Random House, New York, 1960).

2. For a general history of the oil industry in this period, see Harold Williamson, et al., The American Petroleum Industry: The Age of Energy, 1899-1959 (Evanston, Northwestern University Press, 1963), pp. 167-205. For events in California, see Gerald White, Formative Years

in the Far West (New York, Appleton-Century-Crofts, 1962). For those in Texas, see Joseph A. Pratt, The Growth of a Refining Region (Greenwich, Conn., JAI Press, forthcoming).

3. These are figures from Sam Schurr and Bruce Netshert, Energy in the American Economy, p. 108.

4. Ibid, p. 36.

5. According to the statistics presented in Schurr and Netshert (p. 145), total energy consumption in the United States increased by the staggering total of 73 percent between 1900 and 1910.

6. For an excellent, detailed analysis of regional growth, see Harvey Perloff, et al, Regions, Resources, and Economic Growth (The Johns Hopkins Press, Baltimore, 1960).

7. The National Oil Reporter was an oil industry trade journal that existed briefly in the first years of the twentieth century. It contained detailed reports of all phases of the growth of fuel oil sales, and it regularly listed individual railroads or industrial concerns that had switched from coal to oil. The Library of Congress, Washington, D.C., has a complete collection. For accounts of railroads converting to oil, see, for example, National Oil Reporter, July 18, 1901, p. 8; and September 19, 1901, p. 3. Other descriptions of oil-burning by railroads appear in Oil Investors' Journal, April 3, 1906, pp. 12-13; William B. Phillips, "Fuel Oil in the Southwest" in Transactions of the American Institute of Mining Engineers, vol. XLVIII (New York, 1915). "Petroleum Fuel for Locomotives", Engineering News, August 25, 1892, pp. 174-175; and "Petroleum as Fuel", Power, September, 1902, pp. 20-26.

8. Interview of Henry C. Adams by James Garfield, dated February 28, 1907, File Number 3345, Records of the Bureau of Corporations, Record Group 122, National Archives, Washington, D.C.

9. Sam Schurr and Bruce Netshert, Energy in the American Economy, pp. 115-124.

10. National Oil Reporter, January 1, 1902, pp. 13-14; July 17, 1902, pp. 3-5; November 20, 1902, pp. 3-5; and November 27, 1902, pp. 3-5. For long-run trends in fuel oil use by steamships, as well as by other consumers, see National Industrial Conference Board, Oil

Conservation and Fuel Oil Supply (National Industrial Conference Board, Inc., New York, 1930).

11. The Navy ran extensive tests comparing the relative merits of coal and oil in the first decade of the century. Smaller ships were the first to be converted, but before the beginning of World War I, the Navy had completed the conversion of much of its entire fleet.

12. This estimate is derived from a variety of information contained in the Texaco Archives, White Plains, New York. More systematic statistics on refinery consumption of fuel oil are available for the 1920s. Even then, after concerted efforts to use this fuel more efficiently, refining still used as much as 15 percent of all fuel oil consumed.

13. Sam Schurr and Bruce Netshert, Energy in the American Economy, pp. 125-143.

14. For a contemporary account of this process, see James H. Westcott, Oil: Its Conservation and Waste (Beacon Publishing Company, 1930).

15. National Oil Reporter, December 12, 1901, pp. 8-9; July 17, 1902, pp. 3-5; and July 24, 1902, pp. 3-4. Manufacturer's Record, April 11, 1901, pp. 212-214. Oil Investors' Journal, April 18, 1906, p. 14. Gerald White, Formative Years in the Far West, pp. 314-315.

16. A provocative recent article on the transition from wood to coal--Charles A. Berg, "Process Innovation and Changes in Industrial Energy Use," in Phillip Abelson and Allen Hammond (eds), Energy II: Use, Conservation, and Supply (American Association for the Advancement of Science, Washington, 1978), pp. 3-8--argues that technological advances in several major industries "may have been more important than the relataive price of wood and coal in motivating the conversion of American industry to the use of coal." Berg's argument is persuasive, but it does not necessarily apply to the transition from coal to oil. Coal and wood were very different fuels, and coal produced a far more intense and more easily controlled heat. On the other hand, oil and coal were both fossil fuels that produced fuels of roughly similar quality. Thus price would have become a more important consideration in distinguishing between the two fuels.

17. The price for Texas oil is estimated from the information on oil prices regularly posted in the Oil

Investors' Journal (which later became the Oil and Gas Journal) and from sources in the Texaco Archives and the Bureau of Corporations' Oil Investigation in 1906.

18. Articles on the relative cost of fuel oil and coal and of the cost and the technical problems of converting coal-burning equipment to oil appear in technical journals of this era. See, for example, "Altering Coal Plant for Oil Fuel: Liquid Fuel Burning Installation," Industrial Management, 59, 2 (February, 1920).

19. For the long-run price trends of both oil and coal, see Neal Potter and Francis Christy, Jr., Trends in Natural Resource Commodities (The Johns Hopkins Press, Baltimore, 1962).

20. Early in their histories, Texaco and Gulf Oil built pipelines from their Gulf Coast refineries to the Oklahoma oil fields at a cost of more than seven million dollars each. When Texaco committed itself to build its Oklahoma pipeline in 1906, its total assets were less than eight million dollars. Both companies also invested heavily in tankers in this period.

21. At the turn of the century San Francisco was far larger than the other three cities. But first Los Angeles, and then Houston and Dallas grew much more rapidly than did San Francisco.

22. The Coal Trade, a yearly publication specializing in information about market conditions in the coal industry, contained yearly reports about coal consumption in all of the major cities. For the statistics on San Francisco see Frederick E. Savard, The Coal Trade, 1901, p. 68; and 1905, p. 92.

23. See, example, The Coal Trade, 1908, p. 110.

24. This early concern about the possible exhaustion of oil reserves peaked in the years immediately after World War I. The fact that these expert predictions were quickly followed by vast new discoveries of oil helped to discourage further worry about the adequacy of oil supplies. For a fascinating look at this problem from the perspective of two energy experts in the 1920s, see Chester G. Gilbert and Joseph E. Pogue, America's Power Resources (The Century Co., New York, 1921).

25. Coal strikes could last for months, and the impact could be especially severe on the industrial centers of the

Northeast. In the often intense competition in these areas between Pennsylvania coal and Gulf Coast oil such strikes often introduced otherwise reluctant consumers to the advantages of fuel oil over coal.

26. The Hoover quote is from Chester Gilbert and Joseph Pogue, America's Power Resources, pp. 81-82. For the state of the industry in the 1920s, see Wilton Hamilton and Helen Wright, The Case of Bituminous Coal (The Macmillian Company, New York, 1926).

27. For an excellent history of Standard in this period, see Ralph and Muriel Hidy, Pioneering in Big Business, 1882-1911 (Harper and Brothers, New York, 1955).

28. Most of the information in this section is taken from the records contained in the Texaco Archives at White Plains, New York. For Texaco's history, see Marquis James, The Texaco Story: The First Fifty Years (The Texas Company, 1953); John O. King, Joseph Stephen Cullinan (Vanderbilt University Press, Nashville, 1970); and Joseph A. Pratt, The Growth of a Refining Region.

29. I have developed this argument in much greater detail in an article now being prepared for publication, "The Petroleum Industry in Transaction: Business-Government Relations in the Early Texas Oil Industry." My primary sources of information were the Texaco Archives and the records of the Bureau of Corporations' extensive investigation of the national petroleum industry in 1905 and 1906. (Record Group 122, National Archives).

30. The Texaco Archives include numerous good sources on the development of this Mississippi River trade. These are easily located in the card index at the Texaco Archives under the headings "Mississippi River Trade," "Jung," and "Crusel."

31. The progress of this competition between oil and coal on the East Coast can be observed in the yearly reports about the major coal markets in The Coal Trade.

32. Alfred D. Chandler, Jr., The Visible Hand: The Managerial Revolution in America (The Belknap Press, Cambridge, 1977), pp. 503-512.

33. A good account of these disputes is found in Lorenzo Meyer (Muriel Vasconcellos, translator), Mexico and the United States in the Oil Controversy, 1917-1942 (University of Texas Press, Austin, 1977).

34. John Ise, <u>The United States Oil Policy</u>, (Yale University Press, New Haven, 1926),p. 159. See also, George Ward Stocking, <u>The Oil Industry and the Competitive System: A Study of Waste</u> (Houghton Mifflin Company, New York, 1925).

August W. Giebelhaus

2. Resistance to Long-Term Energy Transition: The Case of Power Alcohol in the 1930s

In the midst of the current national debate over America's energy needs, the issue of "power alcohol" or alcohol used in the internal combustion engine either by itself or in a blend with gasoline, has again raised its head. Proponents have suggested the use of two types of alcohol for motor fuels. Methanol (wood alcohol) is obtained either from the distillation of wood products or by synthesis from carbon monoxide and hydrogen. Ethanol (grain alcohol) is derived by the age-old process of fermentation of plant materials such as corn, wheat, sorghum, or sugar cane, or by synthesis from ethylene gas [1]. Thus, part of the complexity of today's debate involves argument over methanol versus ethanol as well as the technical, economic, and political issues related to the uses of alcohol as motor fuel.

This paper focuses on the ethanol side of the question. Today, as in the past, many of the leading advocates of power alcohol come from the midwestern farm belt and are urging government support of programs to produce ethanol from fermented grain or agricultural waste. This program would have a two-fold purpose. Not only would utilization of a renewable energy source (grain alcohol) reduce oil imports and strengthen our future energy position, but it would provide immediate relief to a discontented farm sector burdened with higher production costs, lower profit margins, and surpluses in key areas of the agricultural market [2].

There is nothing new about recent programs to market "gasohol" (the trade name of the 10% alcohol-gasoline blend now marketed nationally) except that the economics of alcohol motor fuels make more sense in an era of rapidly escalating petroleum prices and diminishing reserves. Although controversy has accompanied each previous attempt to introduce farm-brewed alcohol into the American motor car, cost arguments rather than technical obstacles have defeated power alcohol in the past. During periods of agricultural surplus and/or concern for petroleum supplies, alcohol fuel programs have emerged. In this discussion, I will examine the power alcohol movement during the Great Depression of the 1930s, the most organized previous effort to introduce alcohol as a motor fuel. I will analyze why the movement failed, focusing primarily on the opposition of the petroleum industry and related segments of the motor users' lobby.

Historical Overview

Agitation for alcohol fuels has coincided with low farm prices and pessimistic estimates of America's petroleum reserves. As early as 1906 there were serious efforts to promote industrial alcohol as a fuel. Agricultural leaders and industrial chemists supported legislation removing the federal tax on beverage alcohol in the hope that new markets would develop for alcohol fuel. Farmers sought markets for surplus crops, a group of energetic chemists were enthralled with the prospects of new scientific applications to industry, and both united to combat what appeared to be growing shortages of petroleum supplies [3]. Cheap oil from the new Oklahoma fields and a lack of public acceptance of alcohol contributed to failure at this time.

With the outbreak of World War I, Henry Ford campaigned for the production of alcohol from grain and garbage both as a way to combat gasoline shortages and as a cornerstone of his growing interest in wedding agriculture to industrial science [4]. There were only limited experiments with alcohol fuels during the war, but pessimistic estimates of petroleum reserves after 1918 gave renewed life to the idea. Chemical engineers lobbied to isolate industrial alcohols from alcoholic beverages in the drafting of prohibition enforcement legislation on the grounds that alcohol had great potential for motor fuels "in the future." In an atmosphere of short-term scarcity, advocates also stressed the technical advantages of increased power and antiknock from alcohol blends as compared with straight gasoline [5].

In 1922-23 the Standard Oil Company (New Jersey) marketed a 25% alcohol-gasoline blend in the Baltimore region during a period of high gasoline and unusually low alcohol prices. However, the firm encountered customer dissatisfaction. Although alcohol and gasoline are fully miscible, the alcohol must be anhydrous, with as little water present as possible. Small amounts of water caused separation of the alcohol and gasoline in storage tanks as well as in the gas tanks and carburetors of automobiles. The alcohol also worked as a solvent and quickly clogged fuel lines with sediment it had loosened. Consequently, Standard abandoned the experiment after two years [6].

Power alcohol proved to be impractical in the 1920s for three major reasons. Huge strikes of crude oil in California and in the mid-continent fields provided abundant supplies of oil, facing the industry with the problem of overproduction rather than that of scarcity. Second, improvements on the basic process of thermal cracking provided increased quantity and quality of gasoline. Third, the introduction of tetraethyl lead offered an additive to increase gasoline octane which was cheaper than alcohol. Although the farm sector remained depressed throughout the decade, the abundance of cheap gasoline, and the red tape involved in producing alcohol in the era of prohibition, combined to effectively kill the power alcohol movement [7].

Alcohol and the Great Depression

The election of Franklin D. Roosevelt to the Presidency in 1932 sparked renewed interest in alcohol. The repeal of prohibition promised to give a freer hand to distillers to produce industrial alcohol as well as potable spirits from surplus grain. Alcohol motor fuels represented a solution to the malaise of cornbelt farmers. For alcohol advocates the movement was a panacea; for critics it represented an uneconomic gimmick that would cost the consumer money and at best be only a temporary expedient for the farm community.

Among the pro-alcohol arguments put forth in 1933 was the observation that several foreign nations had been successfully utilizing alcohol blends with or without government support for many years. Without exception, however, the nations that had adopted compulsory alcohol blending were net importers of petroleum, and promoted alcohol as part of a policy of increased fuel self-sufficiency [8]. Alcohol supporters in this country introduced similar nationalistic arguments. Although the alcohol program's presumed farm relief benefits occupied center stage, the program's champions related alcohol fuel to longer-term national policy, citing the fifth annual report of the Federal Oil Conservation Board (1932), which emphasized the uncertainties of future oil reserves. Stressing the fact that petroleum was an "exhaustible" resource, the FOCB had estimated that, at current rates of production, most known oil resources would be depleted within ten to twelve years. To further bolster their argument for supporting alcohol fuels, proponents also cited the military importance of alcohol in the manufacture of munitions--distilleries located in the farm belt rather than the more vulnerable east coast would be a valuable asset in the event of war [9].

The revival of the alcohol fuel idea began in the region around Peoria, Illinois, long a distilling center. In January 1933, Paul Beshers, a canning factory chemist, lobbied Congress to support a plan requiring that all motor fuel be a blend of 10% ethyl alcohol made from farm products. Local organizations began a campaign of letter writing, held public meetings, and distributed pamphlet literature touting alcohol fuel as a "plan for national economic recovery." Lame-duck Congressman William E. Hall of Peoria introduced a bill providing for a 10% tax on unblended gasoline, and other proposed legislation called for a national program gradually leading to all motor fuels containing 10% alcohol. These bills became lost in the legislative stalemate of the Hoover-Roosevelt interregnum and in February 1933 the focus of activity shifted to the state houses of Illinois, Iowa, Indiana, Nebraska, and other corn-growing states. National attention focused on legislative hearings held in Iowa on a bill to mandate the use of 10% farm alcohol in all motor fuels in that state. Testimony from both sides of the debate in this hearing highlight most of the issues that continued throughout the 1930s and still are debated in regard to the gasohol movement [10].

A committee from Iowa State College at Ames headed by Drs. Charles E. Friley and Leo M. Christensen gave evidence to show that alcohol-blended fuel was not only as good as straight gasoline, but was far superior. Using anhydrous alcohol (less than 0.5% water) in a 10% alcohol blend, they found no serious problems of separation. The Ames researchers also reported improved antiknock ratings and increased power. They used the 10% alcohol blend with straight gasoline in tests of twelve cars at constant speeds without any carburetor adjustment for the alcohol mixture. The blend gave 5% better mileage at speeds of 20 miles per hour, and 0.6% better mileage at 50 miles per hour [11].

The Ames performance tests appeard to refute what had become one of the basic criticisms of alcohol blends. Since alcohol contains only about half as much energy as an equal amount of gasoline, it appeared logical that a 10% blend would reduce mileage by approximately 3-4% per gallon. Although there has been much controversy over this issue for a number of years, tests indicate that the leaner fuel-air mixture of 10% blends does produce more efficient combustion of the gasoline and can result in increased mileage. The 1933 data also showed that more complete combustion of the alcohol blend also resulted in 40-50% less carbon monoxide emission [12].

L. S. Bachrach, an alcohol manufacturing consultant from New York, led the opposition testimony at the Iowa hearings. His main argument was that power alcohol was economically impractical as a farm relief program. Bachrach estimated that 10% alcohol added to gasoline with corn at 25 cents a bushel would raise the price of fuel 2.25 cents a gallon; at 50 cents a bushel, 3 cents; and at 75 cents a bushel, 3.6 cents. He calculated these prices on the cost of alcohol manufacture alone and did not incorporate charges for blending or distribution. Bachrach concluded that, as a result of their own higher fuel costs, farmers would break even only with corn at 75 cents a bushel, and that, therefore, the program made no sense as an agricultural relief measure [13].

Other witnesses also addressed the impracticality of the scheme. R. E. Rhodes, of the Iowa Motor Club, criticized the loss of tax revenue that would result from a lower demand for higher-priced fuel. Farm journalist Kirk Fox argued that fuel blending would utilize only 3% of Iowa corn production, and motor expert George Granger Brown of the University of Michigan again raised many of the technical criticisms that historically had plagued power alcohol. Among these were the excessive costs in manufacturing anhydrous alcohol; the necessity of adding expensive blending agents; the problem of water absorption leading to fuel separation; the dissolving of shellac on cork floats in carburetors and fuel gauges; the clogging of fuel lines by scale removed by alcohol; and alcohol's corrosive effect on engines [14].

When the heated legislative debate ended, the Iowa bills to establish state-controlled alcohol distilleries and require mandatory 10% alcohol blends went down to defeat. However, this was only the first round in a bitter struggle which would involve farmers, corn products manufacturers, industrial alcohol distillers, and industrial chemists on one side, with the major oil companies, automobile manufacturers, and auto clubs and other consumer groups on the other.

Oil Industry Opposition

The near passage of compulsory power alcohol legislation in Iowa, and the introduction of similar laws in South Dakota, Illinois, and Minnesota, shocked the oil industry into action. The harsh tone of initial individual attacks on the alcohol proposal was reflected in a series of critical, often sarcastic articles which appeared in petroleum trade journals and in the business press during February and March of 1933 [15].

In order to overcome this opposition, the Motor Fuel Alcohol Committee, an organization representing several corn belt groups who supported alcohol blends, began to lobby for a more gradual blending program. This plan called for an amendment to the Revenue Act of 1932 to continue the one cent per gallon federal tax on gasoline blended with alcohol from farm crops, but to increase the tax to two cents per gallon on straight gasoline. Starting January 1, 1935, the tax would be increased to three cents per gallon. To qualify for the lower tax rate, oil companies would only have to blend a minimum of 1% alcohol through December 1, 1934, and thereafter, 5%. Although the MFAC hoped that this gradual introduction of alcohol would be palatable, the organization envisaged that oil companies would use 10% (rather than 1% or 2%) blends in order to take advantage of the high octane and clean burning attributes of the blended fuels [16].

Fearing that the arena was again about to shift to Washington, the oil industry began to coordinate its opposition through its leading trade association, the American Petroleum Institute. The Institite's Industries Committee organized opposition to the pending national legislation. In addition to the argument that alcohol blends would be more costly to the consumer (and also would cut down sales), the API emphasized the potential problem of tax evasion by violators selling straight gasoline at the lower taxed price, and the tremendous expense that the industry would incur in complying with the proposed laws. Refining equipment, storage facilities, and transportation systems, the API argued, would have to be altered to accommodate alcohol blends [17].

The Institute directed much of its criticism at alcohol distillers who, the API charged, were "very active in assisting in promoting this proposal, which is really a scheme to sell the farmer another gold brick, this time at the expense of the motorist." [18] In a high priority memo of April 15, 1933 to its members, the API Industries Committee outlined a plan for action:

> The situation is so critical that action must be taken at once. Make contact immediately with automobile clubs, motor transport interests, and similar organizations in your territory. Acquaint them with the real nature of this scheme and enlist their support in opposing this legislation. Direct the attention of newspaper, trade paper, and automotive

publication editors to alcohol blend legislation, and encourage them to consider the economic advisability. Inform oil companies in your territory of the possible disastrous effects and urge them to do whatever they can to make all the facts known to their contacts, particularly customers. Overlook no opportunity to arouse discussion of the subject particularly of its social and economic phases. Urge individuals and organizations to write to their representatives in Congress, and in state legislatures where such legislation has been introduced, to oppose such legislation. Prompt action is necessary to counteract the aggressive and misleading propaganda of the proponents of the blending schemes. [19]

Reprints of anti-alcohol articles appearing in Business Week, the Wall Street Journal, and Industrial and Engineering Chemistry accompanied this memo (20).

Active oil industry efforts at the state level supplemented the national API campaign. In Pennsylvania, for example, the Associated Petroleum Industries held meetings to coordinate their activities. A list of the Pennsylvania group's ten "Tentative Suggestions for the Organization of a Campaign to Combat Legislative Proposals Requiring Alcohol Admixtures in Motor Fuel" included statewide dissemination of "literature and propaganda," the arranging of local protest meetings, contacting United States Senators and Representatives, working together with other members of the Highway Users' Conference, and cooperating with local Chambers of Commerce and service clubs [21].

Meanwhile, the API distributed copies of printed literature, including the critical testimony given by L. S. Bachrach during the Iowa hearings. On April 21, 1933, Gustav Egloff of the Universal Oil Products Company presented a paper at the semi-annual meeting of the National Petroleum Association in Cleveland. Reproduced and disseminated by the API Oil Industries Committee, Egloff's paper was highly critical of power alcohol. He concluded that alcohol-gasoline blends were inferior to gasoline alone as a fuel, and he listed fifteen specific technical problems [22].

In addition to these technical arguments, Egloff repeated the economic criticisms first highlighted by Bachrach and introduced a "moral" argument which increasingly became a major issue with the oil industry--bootlegging. If alcohol gasoline blends came into general use, the bootlegger would have a ready supply of alcohol. He could simply separate the alcohol from the gasoline by the addition of water and then convert the denatured alcohol into potable moonshine. In somewhat sarcastic language Egloff concluded, "to force the use of alcohol in motor fuel would be to make every filling station and gasoline pump a potential speakeasy." [23]

To further counter the "propaganda" emanating from Peoria and points west, the oil companies used the radio to get their message across to the public. On his evening broadcast of April 25, 1933, popular news commentator Lowell Thomas read a message from his sponsor, the Sun Oil Company, in which he labeled the pending alcohol-blend legislation as another "quack harmful panacea" for the farmer which would have injurious effects for the consumer as well as for the gasoline and motorcar manufacturer [24]. On the Solly Ward program of April 26, sponsor Standard Oil (N. J.) highlighted the bootleg program and stated that the legislation would "make alcoholics out of America's 22,000,000 motor cars" [25]. This disparaging rhetoric was typical of that used by the oil companies in their attacks on the "alky-gas" scheme.

In reply, power alcohol proponents Leo Christensen, Ralph Hixon, and Ellis Fullmer drafted a defense of the program. This monograph, published in 1934, singled out "the several Standard Oil Companies," the Texas Company, Sun Oil, Universal Oil Products Company, the American Automobile Association, and the American Petroleum Institute Industries Committee for their opposition to power alcohol. The authors presented a line by line defense of the technical and economic criticisms made by Egloff and Bachrach, but reserved particular bitterness for the oil industry's injection of the liquor issue into the power alcohol debate. Christensen et al. quoted a Standard Oil (Indiana) circular, which raised the question of water causing separation and enabling bootleggers to easily acquire "drinkable moonshine". The authors argued that it was extremely difficult to make such alcohol potable, and they maintained that the Bureau of Industrial Alcohol did not view the distribution of power alcohol as a threat to the liquor control laws [26].

In response to the oil industry's charge that preferential tax policies were "class legislation" placing burdens on one industry to aid another, the alcohol camp replied that most laws are such and that "it has been the general consensus of opinion that such class legislation is justifiable if the indirect effects result in the greatest good to the greatest number." [27] Arguing that the proposed tax incentive legislation did not compel anyone to sell gasoline mixed with alcohol, the authors claimed that power alcohol would benefit the oil companies. They cited statistics showing that farmers were replacing trucks, tractors, and automobiles with draft animals, and argued that power alcohol would significantly increase oil company sales in the farm belt if only the major companies would go along with this plan of farm relief [28].

The oil industry's all-out blitz to delay and defeat legislation supporting power alcohol was successful. Effective lobbying in Washington and a large publicity campaign were able to overcome the power alcohol movement's efforts in 1933 and 1934. Bills introduced in both the United States Senate and House died in committee. Failure to pass national legislation left power alcohol a fragmented movement in certain midwestern states. In Illinois, farm cooperatives in the Peoria-Pekin region briefly had distributed a blend under the label of "Hi-Ball Fuel," and D. B. Gurney of South Dakota later marketed a brand of power alcohol in several states. However, these and other isolated ventures were limited to the agricultural market and in most cases sales resulted from farm loyalty rather than competitive economics [29].

The Farm Chemurgic Movement

With its legislative program in the doldrums, power alcohol received a major boost in April of 1935 when Francis Patrick Garvan, president of the Chemical Foundation, a nonprofit educational enterprise established for administering patents and fostering the interests of industrial chemistry, announced the meeting of the first Dearborn Conference for the promotion of science, industry, and agriculture. Hosted by Henry and Edsel Ford, the conference convened in Dearborn, Michigan on May 7-8, 1935. Over 300 representatives came together to discuss an alternative to the acreage restriction farm programs of the New Deal. Their program was beautiful in its simplicity. By encouraging industrial uses of agricultural products, they could create vast new markets and bring an end to America's farm problem. They envisaged power alcohol as the cornerstone of a program which also included the industrial uses of soy beans, Southern soft pine wood pulp, and other agricultural products [30].

The Chemical Foundation had emerged after World War I with the avowed purpose of promoting organic chemistry in the United States. The Foundation's strength lay in its ownership of confiscated German patents, many of which had been the property of the giant I. G. Farben combine. Francis Garvan had been Alien Property Custodian after the war, but resigned to head the Chemical Foundation once it had obtained the German patents. Despite some criticism, Garvan and the Foundation survived a stormy period during the Harding Administration until the U. S. Supreme Court upheld the legitimacy of the Foundation's patent ownership [31].

Along with two pioneer researchers in organic chemistry (Dr. Charles Holmes Herty of Savannah, Georgia, and Dr. William J. Hale of the Dow Chemical Corporation), Garvan, a lawyer himself, had led a successful promotional campaign for organic chemistry during the 1920s [32]. In 1934 Hale had published a book, The Farm Chemurgic, in which he coined a word that would later have important propanganda value in the 1930s. Derived from the "Egyptian" word chemi (origin of chemistry) and from the Greek work ergon (work), the phrase "farm chemurgic" literally meant putting chemistry and related sciences to work to unite farm and industry [33].

Hale, an associate of Henry Ford, had developed a particular interest in power alcohol. He convinced Garvan and the Chemical Foundation to publish the Christensen, Hixon, and Fulmer manuscript as part of its "Deserted Village" series. In early 1935 Hale, Christensen, and William Buffum of the Foundation visited Yankton, South Dakota to see D. B. Gurney, whose "House of Gurney" had been marketing power alcohol in five states. From him they learned that one of his chief problems was securing a steady source of alcohol for fuel blending. Soon afterward, Hale, Buffum, and Garvan made contacts with Henry Ford and others interested in the chemurgy idea and the Dearborn Conference resulted [34].

In dramatic fashion on May 7, 1935, the representatives to the first Dearborn Conference met to sign a "Declaration of Dependence upon the Soil and of the Right of Self Maintenance." The document rested at the entrance to Henry Ford's replica of Independence Hall in his Greenfield Village Historical museum. The desk used once belonged to Thomas Jefferson, the ink stand was a replica of the one used in Philadelphia in 1776, and the table and chairs came from Abraham Lincoln's office in Springfield, Illinois [35].

The delegates elected Garvan to serve as Chairman of the conference; they soon decided to form a new organization, the Farm Chemurgic Council, with Garvan as president. The Chemical Foundation guaranteed financial support of the Chemurgic Council for its first three years of existence [36]. Irene DuPont dampened some of the enthusiasm for power alcohol during a general session at the beginning of the Dearborn meeting when he stated that, in his opinion, science would develop a better fuel than gasoline before alcohol blends could become competitive. DuPont proceeded to say that he did not want to put "a frost" on this proposition, and that he thought it wise to have a free and open discussion [37]. DuPont, of course, had competing loyalties. His firm, E. I. DuPont DeNemours, profited from the several patents held by the Chemical Foundation. He and Garvan were friends as well as business associates. However, DuPont was also a member of the board of directors of the Ethyl Corporation, the firm that held a monopoly on tetraethyl lead. Thus, not only did DuPont have financial interests opposed to power alcohol, but he also was part of the gasoline and automobile lobby [38].

Despite this pessimistic note, the Conference devoted its morning session on May 8 to a discussion of power alcohol. Leo Christensen (then representing the Chemical Foundation), George Granger Brown of Michigan (representing the American Petroleum Institute), and William J. Hale presented papers. The arguments, both pro and con, echoed the rhetoric of the preceding two years. However, at the end of his talk Professor Brown read a letter addressed to Garvan from Axtell T. Byles, president of the API, in which Byles offered to jointly sponsor an impartial "economic and technical" investigation of the relative merits of alcohol-gasoline blends and of straight gasoline. The API offered to assume one half of the necessary expense up to $15,000 [39]. At the committee meeting of the Farm Chemurgic Council on June 17, 1935, the Council decided to refuse the API offer on the grounds that "no proof of feasibility is needed." [40] Garvan cited the success of power alcohol in 22 foreign nations and other evidence to indicate that the alcohol program was here to stay. Since he now represented both the Chemical Foundation and the Farm Chemurgic Council, Garvan was prepared to provide the type of support that power alcohol had lacked in 1933.

The Atchison Experiment

On November 14, 1935, Garvan sent a letter to the chief of the Bureau of Chemistry and Soils of the U. S. Department of Agriculture in which he enclosed a comprehensive proposal for "A Plan Coordinating Agriculture, Industry, and Science." In addition to a general appeal for Department of Agriculture cooperation, one of the document's key provisions was a plan for the construction of a power alcohol demonstration plant. Located near sources of agricultural raw material, the plant would consist of a large unit designed to use Jerusalem artichokes or girasole (a high carbohydrate tuber touted by Hale and other chemurgists), and smaller units equipped to use surplus corn and potatoes. The report recommended Omaha as a site for the plant, and listed Kansas City as a second choice. Both areas had good agricultural supplies, had large livestock and dairy industries to test the value of by-product wet feeds, and had a good demand for dry ice, another by-product of the fermentation process. The Chemurgic Council sought financial support from the Department of Agriculture for the million-dollar demonstration plant [41].

No reply came from the Department, and the chemurgists were unable to drum up external support from industry. However, when the Second Dearborn Conference opened in May of 1936, the Chemical Foundation made an important announcement. William W. Buffum, treasurer and general manager of the Foundation, revealed plans for the creation of "America's first power alcohol plant" to be located on the site of the Bailor Manufacturing Company, an industrial distillery in Atchison, Kansas. Originally designed for the manufacture of farm implements, the plant was converted to alcohol production in 1934. The Chemical Foundation loaned $116,000 to Bailor for equipment purchase, and formed the Chemurgic Foundation of Kansas to handle the plant's output. This experimental plant would use Jerusalem artichokes and sweet potatoes, as well as sorghum, corn, potatoes, and grain, for fermentation. The goal was to produce 10,000 gallons of anhydrous alcohol and 32 tons of protein feed a day [42].

Garvan had set the stage for the Atchison announcement with a resounding speech. After first expounding on such topics as the "scientific habit of thought" and the advantage of good arithmetic, the Chemurgic Council president praised power alcohol as a "positive solution to the farm problem." He enlivened his discussion by launching a personal counter-offensive against the oil industry which he punctuated with appropriate exhibits distributed to the audience. Garvan quoted a statement in <u>Colliers</u> magazine by Jersey Standard president Walter Teagle that the United States would eventually run out of oil; Garvan then threw down the gauntlet by stating, "It is only a question of whether you want to cheat yourself, cheat your children, or cheat your grandchildren." [43] He proceeded to answer the critics of power alcohol with a point by point analysis of why blended power alcohol was a perfect fuel.

At this point Garvan pulled out his trump card. He quoted from the advertising literature of the Cleveland Petroleum Products Company Ltd.--the marketers of Cleveland Discol, a one-third alcohol, two-thirds gasoline blend sold in Great Britain--which described Cleveland Discol as a perfect fuel, a characterization with which Garvan expressed total agreement. He then took delight in announcing that Cleveland Petroleum Products Company was owned and operated by Standard

Oil (New Jersey), one of the leaders in the attack on America's power alcohol program. Indeed, Standard had acquired control of Cleveland Petroleum Products in 1932 and had begun to market Cleveland Discol in conjunction with Distillers Company Ltd., a British manufacturer of industrial alcohol. To demonstrate further the apparent hypocrisy of American firms, Garvan also cited advertising brochures from companies who manufactured alcohol-fueled equipment for foreign markets. Among these were International Harvester (who made farm equipment for the Philippines), Studebaker, Whitcomb Industrial Locomotives, and General Motors [44].

Fred A. Eldean, an assistant to president Axtell Byles of the API, had the unenviable task of responding to the papers on industrial alcohol at the Second Dearborn Conference. Eldean maintained that oil, too, was a product of the land, and one that particularly benefited many farmer-leaseholders. He also pointed out, in the spirit of chemurgy, that the petroleum industry was also becoming a chemical industry as a result of advancing petrochemical developments. Eldean then criticized the "alarmists" who were predicting oil depletion in 25 years. To refute this negative view, he cited recent API surveys which showed secure reserves of petroleum for the near and distant future. Criticizing those who were still urging compulsory alcohol legislation in defiance of economic reality, Eldean concluded by stating that the petroleum industry supported those industrial uses of farm crops proven practical on their merits, and not those simply mandated by legislation [45].

The Chemurgic Council's counterattack in 1936 also consisted of support from sympathetic farm journals and other organizations. The Council sent letters in support of power alcohol (enclosing copies of Teagles' Colliers article and of the Cleveland Discol and Studebaker ads) to all the major oil companies. In addition to appeals to the nationalistic theme of energy self-sufficiency, the Council's literature touted the advantages of blended fuel over straight gasoline [46].

The combined activities of power alcohol supporters following the Second Dearborn meeting in May 1936, finally prompted the Department of Agriculture into action. On May 27, Secretary of Agriculture Henry A. Wallace formally replied to Garvan's November 1935 letter by giving support to the Chemurgic Council's "Plan for Coordinating Agriculture, Industry, and Science." On the subject of power alcohol, Wallace supported the construction of a test plant, proclaiming his long-held belief that this step was necessary, but complaining of a lack of funds. Since the Chemical Foundation had already announced its backing of the Atchison experiment, Wallace's financial disclaimer was irrelevant as well as belated [47].

In his study of the Farm Chemurgic Council, Carroll Pursell concludes that politics played a major role in forcing Wallace's hand. Fearing that the Republican party would adopt a strong Chemurgy plank for the 1936 election, the New Deal was now trying to undercut the Republican position by supporting the Council's program. If this was the case, the gesture was futile since neither party really made much of chemurgy in 1936 [48].

The Farm Chemurgic Council publicly disassociated itself with attempts to make power alcohol compulsory by legislation. In letters to oil industry leaders, the Council emphasized that the Atchison plant's purpose was to yield cost data and technical information on alcohol production. This was a tactical maneuver aimed at disarming some of the plant's opoposition and perhaps enlising oil industry support. The Council continued to disseminate literature about the Cleveland Discol situation, however, in an attempt to highlight the oil industry's double standard. The Cities Service Oil Company, for example, had also begun to market an alcohol blend in Britain under the tradename, "Koolmotor," and the Chemurgic Council distributed copies of the product's promotional literature [49].

The Atchison, Kansas plant began operations on October 1, 1936. The plant had a rated capacity of 10,000 gallons of anhydrous ethanol per day to be produced from grains or syrups. (Black strap molasses, not grain, long had been the major raw material for production of industrial alcohol.) After denaturing the alcohol in compliance with federal law, the firm marketed it under the tradename of "Agrol Fluid." [50]

The first six months of plant operations were discouraging: Leo Christensen, the plant's technical director, discussed Agrol's problems at the third Farm Chemurgic Council meeting in Dearborn in 1937. Legal problems relating to the denaturing process, and difficulties in obtaining steady supplies of agricultural raw material from the local farm community were particularly annoying. Both problems impeded the Atchison plant's progress toward full commercial operation. One of the original reasons for using the Bailor plant in Atchison was that it already had a permit to make industrial alcohol. Yet the plant's operators still had to obtain a new permit from the Federal Alcohol Tax Unit. At that point, however, the Agrol people could not reach a satisfactory agreement with the Government on a proper denaturing process. Leo Christensen criticized this government red tape for delaying full commercial production of Agrol [51].

On March 3, 1937, William W. Buffum announced the full commercial operation of the plant under the corporate name of the Atchison Agrol Company. The Chemical Foundation was now the sole owner with Buffum as president, and the Bailor Company was out of the picture. The plant produced Agrol Fluid consisting of 78% anhydrous ethanol, 7% other alcohols, and 15% benzol produced from coal. The latter two ingredients were primarily used as denaturants. These blends initially had a limited marketing area, selling only in Atchison, Kansas and Superior, Nebraska. Christensen estimated the cost of compliance with government regulations at this point to have been $100,000 [52].

Atchison Agrol had now solved most of its legal compliance problems and concentrated on increased sales. The firm distributed an advertising brochure entitled "Agrol--Power Alcohol Made from American Farm Products." The pamphlet summed up most of the performance and agricultural relief arguments that had been current since 1933 [53]. Pumps dispensing the blended gas displayed distinctive globes bearing the Agrol logo, and Atchison marketed three different products. "Agrol 5," containing between 5%, and 7-1/2% Agrol Fluid, advertised an antiknock value equal to "most regular gasolines." "Agrol 15" contained between 10-1/2% and 17-1/2% Agrol Fluid and claimed an octane rating higher than most premium fuels. "Agrol 10," the most popular blend, contained between 7-1/2% and 12-1/2% Agrol and sold at a price usually one cent per gallon above regular gasoline. The firm blended alcohol produced at Atchison at various bulk stations in Kansas, Missouri, Nebraska, Iowa, North and South Dakota, and Minnesota. By the spring of 1938, the height of the Agrol campaign, these blends sold in approximately 2000 service stations in eight midwestern states [54].

Atchison Agrol planned to expand operations by constructing small plants throughout the Midwest. The company promised to construct fermentation plants in any area that could guarantee sales of 3,000 gallons of fluid daily. Part of this change in strategy resulted from the Chemical Foundation's announcement in April 1938 that it was withdrawing financial support from the Chemurgic Council, because the Foundation's principal source of income, the patents it had held on several German chemical processes, had expired the previous summer [55]. Also, the death of Francis Garvan in November 1937 had removed one of power alcohol's staunchest supporters. These developments, however, did not diminish optimism for a booming Agrol program in the spring of 1938 [56].

The location chosen for a second plant was Sioux City, Iowa. Cedar Rapids was under consideration for a third. Demand for Agrol was high in Sioux City, despite the product's still relatively high cost. The local Chamber of Commerce had thrown its support behind the plan, and there were reports that dealers were willing to "accept" a smaller profit margin and sell Agrol blends at the same price as straight gasoline [57].

Power alcohol advocates continued to claim success despite what they characterized as a series of underhanded attacks from the oil industry. In a paper presented at the 1938 Farm Chemurgic Council meeting in Omaha, Paul Beardsley of Sioux City criticized the tactics of the Standard Oil Company (Indiana). He charged that the firm had consistently fought against power alcohol and had spent over $100,000 to defeat pro-alcohol legislation in Iowa, Nebraska, and South Dakota in 1933. More recently, he charged, Standard had threatened the Sioux City Chamber of Commerce that the company would remove Standard's district office from the city if the Chamber persisted in supporting Agrol. Beardsley further claimed that "somebody" called all the retail merchants in town to remind them of the number of customers employed by the major oil companies [58].

Standard firmly denied all charges, indicating that it had made the decision to move its district office well before the power alcohol movement had gained momentum. The firm also corrected another Beardsley charge by pointing out that Jersey Standard, not Standard of Indiana, was involved in Britain with Cleveland Discol. In its published reply, however, the Indiana firm did take the opportunity to attack what it termed the "greatly over-worked" Cleveland Discol issue. Standard pointed out that motor fuel was selling at 31.8 cents a gallon in the United Kingdom and that the British government was subsidizing alcohol production to the tune of 14 cents a gallon as well as exempting it from the 13.8 cents a gallon import duty charged on gasoline. Clearly this was a legal and economic environment different from that in the United States, a major oil producing nation [59].

Agrol executives had also citicized Indiana Standard for pressuring Standard dealers in the Midwest not to market "alky-gas." Agrol maintained that Standard, a large marketer in the agricultural West, had threatened to lift franchises of dealers who sold Agrol, a charge which Standard denied

[60]. The oil industry, and Indiana Standard in particular, were also accused of various unethical activities including the "spreading of lies" about Agrol quality, and the use of rigged "tests" to demonstrate that the gasoline and alcohol in Agrol fuel would readily separate. Agrol charged that traveling "experts" would drive into filling stations and shake up a sample of Agrol fuel in a test tube that they had previously washed. The water in the glass tube naturally caused the mixture to separate. The Agrol people soon instructed its distributors to make sure that only dry tubes were used in similar tests [61].

Agrol's leaders became concerned when imports of Agrol fuel into the Sioux City area declined steadily from a peak of over 30,000 gallons in March 1938 to only 5,000 gallons in July. Initial high demand had been related to Chamber of Commerce enthusiasm and hopes that the Sioux City plant would greatly aid the sagging local economy. But as a result of its own internal financial problems and of the declining interest in Agrol in the Sioux City area (production had never reached the target of 3000 gallons daily), Agrol closed its Sioux City office in May and tabled plans for the construction of the second plant. There had been conflicting Agrol performance reports in Sioux City and no doubt some oil company pressure had hurt demand, but the major factor in Agrol's demise was the small margins that jobbers and dealers made from Agrol sales. By September 1938, Agrol dealers reported that only 5% of their total sales were in alcohol blends [62].

As sales continued to decline, the Agrol Corporation had to reassess its operations. Cost was the major problem that the Atchison plant faced. William Hale's 1936 prediction that 15 cent-a-gallon alcohol could be readily produced did not come to fruition. The best that Atchison could do was 25 cents a gallon, and that was five times the average refinery price of gasoline [63].

The Atchison plant had contracted to buy crops at high prices and was locked in when the market price dropped. Furthermore, although maximum economic hauling distance for raw farm material was usually viewed as 25 miles, the Atchison plant was receiving feedstock from as far as 75 miles. The overall efficiency of the plant during its first year of operation was only 48%; during the second year 63%; and for the third 71.5%. Normal commercial efficiency for industrial alcohol plants was 92%. Atchison Agrol simply could not produce alcohol at a price competitive with straight gasoline [64].

In November 1938, the Chemical Foundation announced that the Kansas plant was closing its doors. The experiment had been a failure. When the Atchison plant closed in November, John Orr Young--a marketing executive who had succeeded Leo Christensen as President of Agrol the previous summer--organized a new corporate entity, the National Agrol Company, for the purpose of licensing whiskey distillers and industrial alcohol producers to manufacture alcohol under the Agrol patents [65].

In December Young wrote to J. Howard Pew, president of the Sun Oil Company, offering him "an inside position" with National Agrol in exchange for Sun's support. Young stated that no other company had been given such an offer. But when Pew turned him down bluntly, stating that the alcohol industry could not function without unwarranted government subsidization, Young presumably contacted other firms. Sun had been a good choice; it was the only major integrated oil company that did not use tetraethyl lead in its gasoline, and the firm had been caught in a squeeze in the octane race that had been going on in the 1930s. However, by the end of 1938, Sun had already invested over $11,000,000 in the development of the Houdry catalytic process, a refining operation that could produce high octane fuel without the addition of either lead or alcohol [66].

Young became frustrated in his attempts to get the Agrol movement going and he resigned in Febraury, 1939, to be succeeded by Nebraska businessman/farmer Frank L. Robinson. In his published letter of resignation Young reported that "the plant has never demonstrated that it could produce from grain a sufficiently low priced product for profitable business." [67] The Agrol project now came to a complete halt. The Reconstruction Finance Corporation did agree to loan $125,000 to reopen the Atchison plant if its managers could raise a matching sum from private sources; they were unable to do so. As of December 1939, the Chemical Foundation had invested over $600,000 in the Atchison project, and auditors' reports indicated operating losses right up to the very end [68].

Final Attempts

In December 1938, the release of an influential government report written by two Department of Agriculture chemists, P. B. Jacobs and H. P. Newton, dealt a further blow to power alcohol. Published as Miscellaneous Bulletin No. 327, "Motor Fuels from Farm Products," this report convincingly argued that it was uneconomic to support a 10% or even 5% alcohol fuel program. From corn at 50 cents a bushel, a low price for 1938, they estimated alcohol could be produced at a cost of 38 cents a gallon. The report criticized power alcohol from almost every angle, but did recommend that the government continue research in the event that future shortages and increased fuel prices would make alcohol blends more competitive with gasoline [69].

The fight was not yet out of the alcohol lobby, and they introduced six new bills into the Congress to exempt alcohol blends from the federal gasoline tax. Two in particular created a stir. Senator Guy Gillette of Iowa reintroduced a bill to exempt all motor fuels containing 7% alcohol made from farm products and Senator Charles Gurney of South Dakota proposed to exempt 10% alcohol blends [70]. Convinced by their own failure to produce cheap alcohol, power alcohol proponents again argued the need for tax incentives to make their program viable. Power alcohol still made sense to them in terms of agricultural relief and long-term renewable energy. Nebraska and South Dakota had passed legislation favoring power alcohol and once again there were hearings being held in Illinois and Iowa [71].

The parade of witnesses before the Senate Sub-Committee on Finance in May 1939 again covered most of the familiar technical and economic arguments. Conflicting testimony of the performance of alcohol-gasoline mixtures still muddied the waters, but even the American Petroleum Institite did not seriously argue that blends could not be successfully used in America's automobiles. Rather, its main criticism centered on cost, and the ethics of subsidizing one industry, alcohol, at the expense of another [72].

One witness's testimony particularly aroused the oil industry. On May 29, P. B. Jacobs of the Department of Agriculture announced the construction of a 300-500 gallon-per-day pilot alcohol plant at the Peoria, Illinois laboratory of the Agriculture Department. The Peoria laboratory was one of four agricultural research labs authorized by Section 202 of the Agriculture Adjustment Act of 1938 [73].

The American Petroleum Institute's Committee on Motor Fuels mobilized to counteract this latest threat in 1939 and 1940. Conger Reynolds, chairman of the API Committee on Motor Fuels, presented another comprehensive indictment of power alcohol at the Chicago meeting of the API on November 14, 1939. The Institute circulated this paper along with other anti-alcohol literature. A highly sarcastic pamphlet entitled "How the Alky-Gas Scheme Would Work," attempted to demonstrate through simplistic reasoning and cartoon figures why it would be more economical to purchase surplus corn from the farmer and burn it on a bonfire than it would be to go through the added expense of fermentation, distillation, and blending with gasoline. Comparing their own figures, the API pamphlet concluded (along with appropriate drawings of cuckoo birds) that "Isn't the alky-gas scheme the crazier of the two?" [74] The combined effects of the oil industry campaign and other readily available negative cost data effectively killed the 1939 and 1940 tax reform bills in committee.

There were to be other flurries of activity in 1940 and 1941, but with war on the horizon the market for farm products improved significantly. The API Committee on Motor Fuels took delight in widely distributing a letter from President Franklin D. Roosevelt to the Speaker of the California Legislature in which he stated that,

> While it is true that a number of foreign countries process agricultural materials for the production of alcohol as a motor fuel, it is equally true that the motor fuel economy of countries possessing no petroleum resources is very different from such economy in the United States. It has never been established in this country that the conversion of agricultural products into motor fuel is economically feasible or necessary for national defense [75].

The President went on to say that no further action on "this contentious subject" should be taken until the Peoria plant was completed and data made available.

The federal government did eventually support alcohol production during World War II, and it even reopened the Atchison plant. However, what was needed then was not power alcohol, but industrial alcohol for munitions manufacture and, after 1942, for synthetic rubber production. Alcohol was a key raw material in the production of both styrene and butadiene, the two substances needed to make strategicallly important buna-s rubber. The alcohol and petroleum industries locked horns over their competing processes for the production of butadiene, but with the end of the agricultural depression, the main argument for subsidizing power alcohol had gone [76].

Conclusion

Most of the arguments against power alcohol in the 1930s are being heard again today in opposition to "gasohol." The oil companies repeatedly charged that alcohol blends corroded engines, that they would separate in storage and in use, and that since alcohol contained fewer BTUs than gasoline, blends inevitably meant a loss of power and mileage. The oil industry opposed preferential legislation to make alcohol more competitive with gasoline, and countered the argument of imminent petroleum shortages by citing API studies indicating secure and abundant petroleum reserves. Just as today, there were also charges from several quarters that it was fundamentally wrong to burn food in engines. In the final analysis, however, the most important issue was economics. Distillers could not produce alcohol cheaply enough to make blends competitive with straight gasoline in the absence of tax incentives. Even without organized oil industry opposition, power alcohol would not have succeeded.

Power alcohol was doomed from the start. The 1930s were a period of enormous overproduction. In 1931, East Texas oil sold for 10 cents a barrel, and the thrust of industry and government policy for the remainder of the decade was to develop acceptable ways to restrict production. Thus, the oil industry's intense opposition to power alcohol was part of its attempt to improve its own price structure and to solve its internal problems in the New Deal era. But the oil men also had the logic on their side. The Atchison Agrol experiment had proven two things. Alcohol blends were technically possible, but not economically viable. The oil industry had been right in the short term.

Today we no longer enjoy cheap gasoline prices. More important, the public has come to accept pessimistic predictions of rapidly depleting oil reserves more readily that it did in the 1930s, even though there still remains controversy over when our petroleum will run out. A growing dependence on imported oil and further price increases by the OPEC nations make power alcohol more attractive than at any other time in the past. However, just as in the 1930s, its strongest advocates present alcohol motor fuel as a farm relief measure first, and only secondarily as an alternative long-term energy solution.

Current and future policy makers must weigh the advantages of ethanol produced from farm products against those of other competing energy sources. They must also decide whether the production of grain alcohol is the most efficient way to aid a farm sector faced with short-term market disequilibrium. A long-term energy source must prove viable for many years to come. When increased demand arises for agricultural foodstuffs, it seems clearer that pressures to expand food production will far outweigh growing raw material for motor fuel.

One of the important differences between gasohol today and power alcohol in the 1930s, however, is that current projects are focusing research on alcohol production from agricultural waste. If we can develop economical technologies to utilize waste streams rather than food crops, gasohol should prove to be more acceptable. Moreover, while gasoline prices continue to increase rapidly there are indications that the cost of alcohol production may be coming down. A newly developed membrane technology, and laboratory improvements in enzymes used in fermentation, promise greater efficiencies in the manufacture of ethanol from farm products or agricultural waste. Methanol, an already less expensive alcohol for blending purposes, is also being produced more cheaply through a coal-systhesis process.

There is another significant external variable which appears to be playing an increasingly important role-- environmental concerns. Unlike the 1930s, we are today in an era of environmental awareness which has seen EPA drastically cut back the use of tetraethyl lead in fuels and ban other additives used for increasing octane. Primarily on the basis of alcohol's octane-boosting properties, the American Petroleum Institute has recently expressed a definite interest in gasohol, a significant departure from its long-held opposition [77]. This tentative interest is tempered, however, by the API's insistence that alcohol must be cost-effective; something that is not yet the case.

History does not repeat itself, but an awareness of past experience can guide us to make more informed decisions in the present and future. Clearly, the economic environment of the 1930s was fundamentally different from that of the 1970s. Yet the main proponent of power alcohol (the farm lobby) and its major opponent (the oil industry) are the same, and their arguments, both pro and con, are strikingly similar.

References and Notes

1. Earl V. Anderson, "Gasohol: Energy Mountain or Molehill?" Chem. & Eng. News 56, 8 (1978); Robert H. Lindquist, "Alcohols and Motor Fuels: The Promise and the Problems," unpublished paper, Chevron Research Company (1977); Thomas B. Reed, "Questions and Answers on Alcohol and Gasohol Fuels," testimony before the Senate Appropriations Committee (January 31, 1978).

2. Anderson, "Gasohol," The New York Times (May 3, 1978); The Wall Street Journal (July 12, 1978); William A. Scheller, "Agricultural Alcohol in Automotive Fuel-- Nebraska Gasohol," paper presented at the Eighth

National Conference on Wheat Utilization Research, Denver, Colorado (October 10-12, 1973); Al Mavis, Farm Energy Conservation Coordinator, State of Illinois, "Statement Before Senate Energy Committee" (August 7, 1978). Proponents of today's power alcohol from farm crops movement have formed the "National Gasohol Commission" to coordinate their efforts in promoting alcohol blends. At the Commission's Chicago Meeting on November 12-14, 1978 it adopted a comprehensive list of seventeen resolutions which largely focus on the need for federal support of gasohol.

3. Conger Reynolds, "The Alcohol-Gasoline Proposal," paper presented at the Twentieth Annual Meeting of the American Petroleum Institute (November 14, 1939); Power Alcohol: History and Analysis, (American Petroleum Institute, New York, 1940), pp. 1-4; John Ware Lincoln, Methanol and Other Ways Around the Gas Pump, (Garden Way Publishing, Charlotte, Vermont, 1976), pp. 92-94; William A. Scheler, "Tests on Unleaded Gasoline Containing 10% Ethanol--Nebraska GASOHOL," paper presented at the International Symposium on Alcohol Fuel Technology--Methanol and Ethanol, Wolfsburg, Federal Rupublic of Germany (November 21-23, 1977); B. R. Tunison, "The Future of Industrial Alcohols," J. of Ind. and Eng. Chem. 12, 370 (1920).

4. Reynold Millard Wik, "Henry Ford's Science and Technology for Rural America,"Tech. and Cult. 3, 248 (1962).

5. M. C. Whitaker to Charles Holmes Herty, June 11, 1919; Whitaker to Hon. Andrew J. Volstead, June 11, 1919 (copy), chronological file, Charles Holmes Herty Papers, Robert Woodruff Library for Advanced Studies, Emory University, Atlanta, Georgia (hereafter cited as Herty Papers); New York World (June 7, 1919); Tunison, "Future of Industrial Alcohols"; Ralph C. Hawley, "The Forests of the United States as a Source of Liquid Fuel Supply," J. of Ind. and Eng. Chem. 13, 1059 (1921).

6. Power Alcohol: History and Analysis, p. 4; "Analysis of Technical Aspects of Alcohol Gasoline Blends," API Spec. Tech. Comm. (April 10, 1933), J. Howard Pew Presidential Papers, Accession #1317, The Sun Oil Collection, Eleutherian Mills Historical Library, Wilmington, Delaware (hereinafter cited as Pew Papers), Box 52.

7. Power Alcohol: History and Analysis, pp. 4-5.

8. Leo M. Christensen, Ralph M. Hixon, and Ellis J. Fullmer, Power Alcohol and Farm Relief, The Deserted Village Series, No. 3 (The Chemical Foundation, New York, 1934), pp. 99-113; "Alcohol Gasoline Blends" (New York: Am. Pet. Ind Comm., 1933), Pew Papers, Box 52; "Permanent Farm Relief Through the Use of Alcohol Blends in Motor Fuel," Motor Fuel Alcohol Comm. (May 9, 1933), Pew Papers, Box 52.

9. "Permanent Farm Relief Through Use of Alcohol Blends," pp. 12-13; L. C. Snider and B. T. Brooks, Petroleum Shortage and Its Alleviation, Deserted Village Series, No. 6 (The Chemical Foundation, New York, 1935), pp. 3-12.

10. "Alcogas," Bus. Wk. (February 8, 1933), p. 9; "Farm-Brewed Fuel," Bus. Wk. (March 15, 1933) pp. 14-16; "Iowa Alcohol-Gasoline Proposal Tabled Temporarily as Idea Grips West," Nat. Pet. News 25, 11 (March 1, 1933).

11. "Farm-Brewed Fuel," pp. 15-16.

12. Ibid.; "Permanent Farm Relief Through Use of Alcohol Blends," pp. 5-7; Reed, "Questions and Answers."

13. "Iowa Alcohol-Gasoline Proposal Tabled," p. 11; L. S. Bachrach, "Facts and Arguments Presented Against Alcohol-Gasoline Motor Fuel Legislation at Hearing Before Iowa State Senate (February 21, 1933), Pew Papers, Box 52.

14. "Iowa Alcohol-Gasoline Proposal Tabled," p. 11; Joseph Geschelin, "Alcohol Mixing Bills Would Make Motorists Pay Far More for Corn Price Rise," clipping in Pew Papers, Box 52; Leo M. Christensen, "Power Alcohol," Proc. of the Dearborn Conf. of Ag., Indus., and Sci., Dearborn, Mich., May 7 and 8, 1935 (The Chemical Foundation, New York, 1935), pp. 103-105.

15. See "Alcogas," "Farm-Brewed Fuel," and "Iowa Alcohol-Gasoline Proposal Tabled Temporarily," cited in note 10 above.

16. "Permanent Farm Relief Through Use of Alcohol Blends," pp. 1-3.

17. R. P. Anderson, API, to J. Howard Pew, April 10, 1933 (with enclosed memoranda on economic and technical aspects of alcohol blends); B. H. Markham, director, American Petroleum Industries Committee, "Memorandum Re

Alcohol-Gasoline Blends," (April 15, 1933), pp. 1-3, Pew Papers, Box 52.

18. Markham, "Memorandum Re Alcohol-Gasoline Blends," p. 1.

19. Ibid., p. 2.

20. "Alcogas" and Farm-Brewed Fuel" (see note 10 above); D. H. Killeffer, "Facts About Alcohol in Motor Fuel," Ind. & Eng. Chem. (News Edition) 11, 117 (1933).

21. W. M. Irish to J. Howard Pew, April 20, 1933; W. Purves Taylor, Assoc. Pet. Ind. of Penn., to W. M. Irish, April 19, 1933 (copy); Taylor to J. Howard Pew, May 2, 1933, Pew Papers, Box 52.

22. Gustav Egloff, "Alcohol-Gasoline Motor Fuels," pp. 1-6, Pew Papers, Box 52; Christensen, et al., Power Alcohol and Farm Relief, pp. 142-48.

23. Egloff, p. 6.

24. Telegram, J. N. Pew, Jr. to H. D. Collier, Standard Oil Co. of Calif., April 26, 1933; Telegram, Lewis Mars, Ind. Alco. Inst., to J. H. Pew, April 26, 1933; Cong. Everett M. Dirksen to Lowell Thomas, May 8, 1933; Thomas to Dirksen, May 18, 1933; Pew Papers, Box 52.

25. W. C. Teagle to J. Howard Pew, April 26, 1933, Pew Papers, Box 52.

26. Christensen et al., Power Alcohol and Farm Relief, pp. 135-59.

27. Ibid., p. 155.

28. Ibid., p. 157.

29. G. D. Jones, Cleveland Tractor Company, to J. Howard Pew, April 26, 1933, Pew Papers, Box 52; Christy Borth, Pioneers of Plenty: The Story of Chemurgy (Bobbs-Merrill, New York, 1939), pp. 74-75.

30. Borth, Pioneers of Plenty, p. 76; Carroll W. Pursell, Jr., "The Farm Chemurgic Council and the United States Department of Agriculture, 1935-1939," Isis 60, 308 (1969); "For Farm and Factory," Time (May 20, 1935), pp. 58-60; "List of Registrants for the Dearborn Conference," May 7, 1935; Invitation (to Charles Herty); Herty Papers, Subject File: "Farm Chemurgy."

31. Pursell, "Farm Chemurgic Council," pp. 307-8; Joseph Borkin, The Crime and Punishment of I. G. Farben (The Free Press, New York, 1978), pp. 30-31, 168-81; Borth, Pioneers of Plenty, pp. 150-53.

32. Carl B. Fritsche, "The Trilogy of Chemurgic Progress," Chemurg. Dig. 19, 5 (1960); William J. Hale, "The Origin of the Farm Chemurgic Movement," Monsanto Curr. Events, clipping in Herty Papers, Subject File: "Farm Chemurgy"; Borth, Pioneers of Plenty, pp. 153-55.

33. Borth, Pioneers of Plenty, pp. 22-24.

34. Ibid., pp. 74-75. Hale later published a second volume devoted to his growing interest in power alcohol-- Prosperity Beckons: Dawn of the Alcohol Era (Stratford, Boston, 1936).

35. Pursell, "Farm Chemurgic Council," p. 309; "For Farm and Factory," Time, p. 58; Copy of Program Ceremony and Declaration in Herty Papers, Subject File: "Farm Chemurgy, Dearborn Conferences."

36. Pursell, "Farm Chemurgic Council," p. 309; Fritsche, "Trilogy of Chemurgic Progress," p. 6.

37. Irenee DuPont, "The Catalyst for Accomplishment," Proc. of the Dearborn Conf., pp. 88-89.

38. Borkin, I. G. Farben, pp. 30-31; Alfred D. Chandler, Jr. and Stephen Salsbury, Pierre S. Du Pont and the Making of the Modern Corporation (Harper and Row, New York, 1971), pp. 523, 570-71.

39. "Morning Session," May 8, 1935, Proc. of the Dearborn Conf., pp. 96-130.

40. "Farm Chemurgic Council, Minutes of Permanent Organization," Committee Meeting, Chicago (June 17, 1935), p. 3, Herty Papers, Subject File: "Farm Chemurgy."

41. Pursell, "Farm Chemurgic Council," pp. 312-13; "A Plan Coordinating Agriculture, Industry, and Science," (November 14, 1935), pp. 30-38, Herty Papers, Subject File: "Farm Chemurgy."

42. W. W. Buffum, "America's First Power Alcohol Plant," Proc. of the Second Dearborn Conf. of Agr., Ind., and Sci. May 12-14, 1936 (New York: The Chemical Foundation, 1936), pp. 108-11; "Business Session of the

Governing Board of the Council," (May 14, 1936), p. 81, Herty Papers, Subject File: "Farm Chemurgy."

43. Francis P. Garvan, "Scientific Method of Thought in our National Problems," Proc. of the Second Dearborn Conf. pp. 73-107; "Chemurgy: Second Dearborn Meeting Fosters Back-To-Farm Movement by Hitching Plow to Industry," Newsweek (May 16, 1936), pp. 32-33; "Chemurgicians," Time (May 25, 1936), pp. 73-74.

44. Garvan, "Scientific Method of Thought," pp. 90-107.

45. Proc. of the Second Dearborn Conf., pp. 131-38; In April 1936 the American Chemical Society sponsored a "Symposium on Motor Fuels" at its annual meeting in Kansas City, Mo. The meeting was an attempt to answer many of the technical and scientific questions that still surrounded power alcohol. The four papers presented, however, again offered conflicting evidence on fuel consumption, engine performance, separation difficulties, and cost. See Gustav Egloff and J. C. Morrell, "Alcohol-Gasoline as a Motor Fuel"; Leo M. Christensen, "Alcohol-Gasoline Blends"; L. C. Lichty and E. J. Ziurys, "Engine Performance with Gasoline and Alcohol;" and Oscar C. Bridgeman, "Utilization of Ethanol-Gasoline Blends as Motor Fuels"; Ind. & Eng. Chem. 28, 1080 (1936).

46. Oswald Wilson, Western Irrigation, to J. Howard Pugh (sic), May 16, 1936; Wilson to Rush M. Blodgett, Oil Producers Agency of California, May 4, 1936 (copy); Carl Fritsche, Farm Chemurg. Council, to J. Howard Pew, August 3, 1936, Pew Papers, Box 52.

47. "Cooperation in Chemurgic Research is Assured by Secretary Wallace," U. S. Dept. of Ag. Press Release (May 27, 1936), Herty Papers, Subject File: "Farm Chemurgy."

48. Pursell, "Farm Chemurgic Council," pp. 313-14.

49. Fritsche to Pew, August 3, 1936; "Koolmotor" pamphlet; Pew Papers, Box 52.

50. Leo M. Christensen, "Agrol-Scientific Aspects," Farm Chemurg. J. 1, 134 (1937).

51. Ibid., pp. 139-41; Leo M. Christensen, "Practical Problems of Converting Farm Crops into Alcohol," paper presented at the Midwestern Conference of Agriculture, Industry, and Science (March 9-10, 1937), pp. 51-59.

52. Farm Chemurg. Council News Release, (March 3, 1938), Pew Papers, Box 52; "2000 Agrol Stations," <u>Farm Chemurg. J.</u> 1, 75 (1938); Christensen, "Practical Problems," pp. 53-54.

53. Agrol brochure, Pew Papers, Box 52.

54. Christensen, "Agrol--Scientific Aspects," p. 138, "2000 Agrol Stations," p. 75; Borth, Pioneers of Plenty, pp. 168-69.

55. M. G. Voorhis, "Chemurgy Aims at Alky Plants for Farm Belt," <u>Nat. Pet. News</u> 30, 25 (May 11, 1938).

56. Borth, <u>Pioneers of Plenty</u>, pp. 147, 168-69; "2000 Agrol Stations," p. 75.

57. Van Voorhis, "High Pressuring for Agrol Plants," pp. 25-26.

58. "Comments by Standard Oil Company (Indiana) on a Speech of Mr. Paul Beardsley--and Statement of Attitude Regarding the Power Alcohol," (May 4, 1938), pp. 1-12, Herty Papers, Subject File: "Chemurgy."

59. Ibid.

60. "Standard Oil Company (Indiana) Denies Charge of Fighting 'Alky-Gas'," <u>Nat. Pet. News</u> 30, 39 (July 27, 1938); "Dealers May Sell 'Alky Gas' if They Choose Says Standard," <u>Nat. Pet. News</u> 30, 39 (July 27, 1938).

61. Borth, <u>Pioneers of Plenty</u>, pp. 169-70.

62. E. L. Barranger, "Agrol Movement Collapses," <u>Nat. Pet. News</u> 30, 25 (Sept. 28, 1938); Barringer, "Agrol is Political Issue in Iowa Campaign," <u>Nat. Pet. News</u> 30, 21 (Oct. 5, 1938).

63. <u>Power Alcohol: History and Analysis</u>, pp. 12-13.

64. Ibid.; Herman Frederick Willkie and Paul John Kolachov, <u>Food for Thought</u>, (Indiana Farm Bureau, Indianapolis, 1942), pp. 11-13.

65. "Chemical Foundation Had Invested More than $600,000 in 'Alky-Gas'," <u>Nat. Pet. News</u> 31, 35 (June 21, 1938).

66. John Orr Young to J. Howard Pew, December 22, 1938; Pew to Young, December 23, 1938; Pew Papers, Box 52.

67. "'Alky-Gas' Plant Seeks Santa Claus," p. 36; Power Alcohol: History and Analysis, pp. 12-13.

68. "'Alky-Gas' Plant Seeks Santa Claus," p. 35; "Chemical Foundation Has Invested More than $600,000 in 'Alky-Gas,'" p. 15.

69. "Farm Crop Alcohol is Found Impracticable as Motor Fuel," Nat. Pet. News 30, 8 (Dec. 28, 1938) P. Burke Jacobs, "Alcohol From Farm Products," Ind. and Eng. Chem. 31, 162 (1939).

70. "Government Experts Oppose 'Alky-Gas' Subsidy," Nat. Pet. News 31, 22 (May 31, 1939).

71. Power Alcohol: History and Analysis, p. 7.

72. "Government Experts Oppose 'Alky-Gas' Subsidy," pp. 26-27.

73. "Government Will Build Pilot Plant at Peoria to Test Costs of Power Alcohol," Nat. Pet. News 31, 22 (May 31, 1939); Pursell, "Farm Chemurgic Council," p. 315; "Government Reveals" 'Alky-Gas' Research Plans, Nat. Pet. News 31, 37 (Sept. 13, 1939).

74. "Warns of New Efforts to Legislate Farm Alcohol into Use as Motor-Fuel," Nat. Pet. News 31, 46 (Nov. 15, 1939); Arnold R. Daum, A.P.I., to "Oil Industry Executives," January 23, 1940; Conger Reynolds, "The Alcohol-Gasoline Proposal"; pamphlet, "How the Alky-Gas Scheme Would Work"; Pew Papers, Box 52.

75. Franklin D. Roosevelt to Gordon H. Garland, June 3, 1941 (copy); API Committee on Motor Fuels Press Release: "F.D.R. Declares Alcohol-Gas No Aid to Defense," (July 30, 1941); Pew Papers, Box 52.

76. James B. Conant, Karl T. Compton, and Bernard M. Baruch (Chairman), "Report of the Rubber Survey Committee" (September 10, 1942), Pew Papers, Box 79; Borkin, I.G. Farben, pp. 77-94.

77. Energy Resources & Tech. 6, 481 (1978).

Arnold Krammer

3. An Attempt at Transition: The Bureau of Mines Synthetic Fuel Project at Louisiana, Missouri

> The increasing demand for gasoline and oil and the rising cost of finding new petroleum, coupled with America's growing dependence on imports and the unsettled international situation, have continued to emphasize the importance of [synthetic liquid fuel]....Supplementing petroleum with synthetic liquid fuels will not only conserve this Nation's petroleum reserves, but also will bring into greater use its tremendous reserves of coal....[1]

This statement, with its emphasis on the soaring demand for petroleum and its caution against dependence on oil imports from unsettled international suppliers, might well have been made in reaction to the latest price increase by the OPEC nations, or in the wake of political disruption in Iran--but it wasn't. These prophetic remarks were made by the Federal Bureau of Mines more than a quarter of a century ago, in its continual campaign to alert the government to the imperative need for a massive program of synthetic fuel production. For just a moment--a short, four-year period out of nearly sixty years of campaigning--the federal government allowed the Bureau of Mines to embark upon a substantial program of coal-to-oil synthetic fuel development. At the program's heart was a pilot project known as the Louisiana, Missouri Demonstration Plant (located in the town of "Louisiana" in the state of Missouri), and it lasted only from 1949 to 1953. During this brief period, however, the Louisiana plant tested every major coal conversion process, engineering technique, and a variety of American coal. That it was able to produce gasoline from coal for an estimated and astonishing 1.6¢ per gallon only adds mystery to the sudden and unexplained closing of the project in June 1953. Its beginning, however, is far more clear, and reaches back to Germany at the end of the First World War.

Early Synthetic Fuel Research

Numbed by her defeat in the cataclysmic world war, and casting about desperately for an explanation, Germany found herself in agreement with Lord Curzon's dictum that the winning side had floated to victory on a sea of oil. This was a bitter lesson which the Germans hoped to avoid relearning in the future. Consequently, a number of German chemists, building on their earlier successful research in synthetic aniline dyes, ammonia, and nitrogen, proceeded to attack the problem of converting Germany's vast resources of coal to the otherwise unavailable petroleum products. By the mid-1920s, they had developed two major processes. The first, developed by Friedrich Bergius (later to win the Nobel Prize for his work), forced a mixture of powdered coal, recycled oil, and a catalyst into a high-pressure vessel filled with hydrogen, where the coal was eventually liquefied. The resulting product was then separated into gasoline, middle oil, and heavy oil; the final products were gasoline and diesel. About 4 or 5 tons of coal were required to make a ton of gasoline, including the coal required to produce the power, steam, and hydrogen used in the process.

In the second process, discovered by two German chemists, Franz Fischer and Hans Tropsch, the powdered coal was broken up by superheated steam to produce a mixture of hydrogen and carbon monoxide gases. This gas mixture, after purifiction to remove all sulfur compounds, was passed over a metal catalyst, under lower temperatures and pressure than required in the Bergius process, to produce a low-octane gasoline, a high-grade diesel oil, and wax which could be further processed into lubricating oils. Like the Bergius direct hydrogenation process, about 4 or 5 tons of coal are required to produce a ton of petroleum product. Whatever the advantages of either process, they both enabled the Germans to obtain gasoline from a coal mine [2].

However, the production of synthetic fuel required acres of pipes, ovens, conveyor belts, and compressors capable of handling pressures from 3,000 to 10,500 lbs. per square inch. This was no operation for small, independent factories, since the capitalization could easily run into the tens of millions of Reichsmarks. Ruhrchemie bought the patent rights to the basic Fischer-Tropsch process and I.G. Farbenindustrie, in turn, bought the Bergius patents, and built the first hydrogenation plant at Leuna in 1927. By the outbreak of war in 1939, fourteen large hydrogenation and Fischer-Tropsch plants were in full operation, with six additional plants under construction. Their combined production of synthetic oil from brown coal of 1,200,000 metric tons (or some 19 million bbls.) per year provided about one-third of the total fuel capacity with which Nazi Germany went to war.

The English also made substantial progress during the 1930s in the development of synthetic liquid fuels from coal. Like Germany, England had few known natural petroleum reserves, and consequently moved into the area of coal conversion far earlier than did the United States. After a thorough investigation of the German direct hydrogenation and Fischer-Tropsch processes (the so-called Falmouth Report), the government underwrote the construction in 1935 of a massive $40 million synthetic liquid fuels plant. Larger than any similar installation in the Third Reich, the Imperial Chemical Industries plant at Billingham, England, was a showcase operation which attracted hordes of visiting scientists from around the world. It employed about 2,000 technicians to operate the plant; required some 4,000 miners and representatives of dependent industries; and produced between 100,000 and 150,000 tons of gasoline per year from coal. The cost per gallon--as calculated by American oil companies who were admittedly wary of the British venture--came to about 18 cents per gallon. Summarizing the reaction of the oil industry to the Billingham plant (and to the looming specter of synthetic liquid fuels in general), Dr. Gustav Egloff of Universal Oil Products Company, a frequent spokesman for the petroleum oligopoly, conceded before a Congressional subcommittee that while the British installation "is a triumph of the first magnitude...from the oil standpoint, or from the competitive fuels standpoint, it is terrifically expensive...." [3]

The French were also active in the development of synthetic liquid fuels from coal. As early as 1922, experiments by E.A. Prudhomme produced water gas from lignite; the gas was then passed over a nickel-cobalt catalyst to produce small quantities of gasoline. Attracted by these experiments for both patriotic and economic reasons, the brilliant French entrepreneur-scientist, Eugene Houdry, in 1927 developed a revolutionary catalytic (as opposed to thermal) process for cracking crude petroleum to produce high octane motor fuel. In 1935, a semicommercial plant, designed to operate at 300 atmospheres, was built in Lievin, France, by Compagnie Francaise des Essences Synthetiques. The installation operated successfully through World War II and was finally dismantled in 1964 [4].

Belgium was also interested in synthetic liquid fuel, from the early experiments on coal conversion at Leuven Katholieke Universiteit in 1924, to the eventual formation of a giant, four-party combine--Union Chimique Belge (process); Geomines (mining); Société Générale de Belgique (finance group); and Comité Special du Katanga (government agency)--to produce synthetic fuels in the Belgian Congo.

Even Asia--not generally considered, in the vanguard of industrial development--had made significant strides in synthetic fuels. In Fushun, Manchuria (of all places), the Central Research Laboratories of the Southern Manchuria Railway Company had begun the experimental hydrogenation of bituminous coal as early as 1928, and in 1939 the Japanese government subsidized the construction there of a 200-atmosphere, semicommercial plant. The Fushun installation produced 15,000 metric tons of oil per year, through 1945. A second plant, also underwritten by Japan, was constructed at Agochi, Korea, by the Korean Nitrogen Fertilizer Company, though the plant was never operated successfully and was converted to produce methanol during World War II [5].

In the United States during the same period, research on synthetic fuel had been nearly nonexistent--and with good reason. America rode through the 1920s and 1930s on a sea of natural petroleum. The number of American petroleum refineries, for example, jumped from an already-impressive 289 in 1919 to 427 in 1929. Production soared from 300 million barrels to 857 million barrels a year during the same period [6]. Production continued to increase throughout the 1930s, reaching one billion barrels in 1939 and nearly a billion and a half by 1941. America went to war as the producer of a whopping 60 percent of the world's output [7]. Moreover, this vast scale of production was achieved without any real consideration of synthetics. There was obviously an abundant, if not unending, supply of natural petroleum.

What little work in synthetic fuels did occur was limited to the first halting steps taken by the Fixed Nitrogen Research Laboratory of the Department of Agriculture at the end of World War I. For the next 20 years this small laboratory provided a steady stream of information on the physics, chemistry, and basic engineering involved in the synthesis of ammonia--an important stage in the future production of synthetic oil. Another step in America's slow journey toward coal conversion occurred with the convocation of the First International Conference on Bituminous Coal in Pittsburgh in 1926. Dr. Thomas S. Baker, then president of the Carnegie Institute of Technology, called the meeting in hopes of interesting America's fuel industries in the pioneering developments achieved by synthetic fuel experts from throughout the world. Friedrich Bergius and Franz Fischer presented papers at the conference, as did George Patart, the French developer of synthetic methanol from carbon monoxide and hydrogen (known as "water gas"). Though the papers presented by these three luminaries revealed to chemists the new horizons in applying catalysis and high-pressure reactions; to oil technologists the new methods for increasing gasoline yields from heavy crude petroleum; and to the coal industry the great possibilities for providing liquid fuels, the results were negligible. Natural petroleum was too abundant for America to be concerned with synthetic fuel from coal.

Not until 1929 did the first significant step toward synthetic fuel in the U.S. occur. Standard Oil of New Jersey--already deeply involved in a clandestine partnership with I.G. Farben which would continue through the 1930s until the outbreak of war--acquired the American rights to the I.G. Farben patents, and made them available to American petroleum companies through its Hydro Patents Company [8]. Several commercial plants actually were built, though their purpose was the hydrogenation of crude oil to produce special items; nothing was done with coal. By 1930 the major oil fields of East Texas and Oklahoma were coming in; the reserve/production ratio of petroleum had increased from 6.5 in 1925 to 15 in 1930; and America was virtually awash in a sea of oil [9].

The U. S. Bureau of Mines

Experimental work on synthetic fuel continued sporadically at some of the small research outposts of the embryonic Federal Bureau of Mines. When the Bureau was created in 1910, it inherited the government's only research program on coal combustion which had begun in the Federal Fuel Testing Laboratory operated at the St. Louis, Missouri Exposition in 1903. During its first decade, the Bureau established eleven additional experiment stations: at Pittsburgh, Pennsylvania; Urbana, Illinois; Salt Lake City, Utah; Golden, Colorado; Berkeley, California; Tucson, Arizona; Seattle, Washington; Fairbanks, Alaska; Columbus, Ohio; Minneapolis, Minnesota; and Bartlesville, Oklahoma [10]. None, however, were initially concerned with anything but problems of local mining and safety.

From 1928 to 1930 the Physical Chemistry Section of the Bureau's Coal Division first conducted some bench-scale experiments on the production of gasoline from water gas by the Fischer-Tropsch process, but a lack of funds soon forced an end to the research. Finally, in 1935, unable to jar the nation from its lethargy, and though still without sufficient funds, the Bureau vowed "to prepare for the time when gradual exhaustion of petroleum resources would require supplementing the growing demand for motor fuel with gasoline and diesel oil from coal." [11] The Bureau embarked on a long-range program of fundamental research to evaluate foreign work on synthetic liquid fuels, as well as to obtain data on the oil yields and operating difficulties of the various kinds of American coals and lignite. The results of such research, it was hoped, would be most useful to American industry in helping it select the most suitable coals when the commercial need arrived; also, better understanding of the chemistry of coal hydrogenation was expected to reduce the cost of this process.

To accomplish these tasks, the Bureau installed a small, continuous unit at its Central Experiment Station at Pittsburgh, capable of hydrogenating 100 pounds of coal every 24 hours. Based on earlier work by the Fuel Research Station of Great Britain, and the Fuel Laboratories of the Mining Department of Canada, the Bureau's project plodded through the next decade with little public notice, testing and cataloging every manner of local coal, catalyst, and process variant, and summarizing its findings in a steady stream of technical reports. Despite the significant advances in technology, and the volume of new information, the results were somewhat disappointing.

Industry, never keenly interested in the relatively expensive production of synthetic oil, now saw its cost-estimates confirmed. According to the Bureau's own figures, the cost of German synthetic liquid fuel by either hydrogenation or gas synthesis fell in the range of between 20 and 30 cents per U.S. gallon [12]. Compared to Germany's reported yields of 140 gallons of oil per ton of Upper Silesian coal, the Bureau estimated that typical American coals yielded only 27.2 gallons per ton of North Dakota lignite, 52.5 gallons from Pittsburgh bituminous, and 58.2 gallons from Utah bituminous--yields corresponding to a disappointing 8.5 to 18.2 per cent by weight of coal used [13]. More to the point, gasoline produced from American coal could range in price from 19.2¢ to 22.6¢ per gallon (total cost, including normal overhead and 10% depreciation), as opposed to only 5.3¢ per gallon of crude (at a nostalgic $1.20 per bbl. and using thermal-cracking techniques) [14]. America's oil companies were understandably unimpressed. Until World War II.

The Impact of World War II

The nature of the new war brought petroleum into sharp focus. The blitzkrieg conquest of Poland, with the emphasis on fighter planes and hordes of tanks, foreshadowed the role that petroleum was to play in the conflict. In World War I, for example, the U. S. military and all its Allies used less than 39,000 barrels (or 1,639,000 gallons) of gasoline daily; in World War II, daily consumption averaged 800,000 barrels (or 33,600,000 gallons)! Indeed, during a period of seven weeks in June and July 1944, the U. S. Fifth Fleet alone used 630,000,000 gallons of fuel oil; and during one month of attacks on Japanese shipping and installations in 1944, the Far Eastern Air Force burned 143,257,000 gallons of aviation fuel. Each sortie by the Ninth Air Force in its daily bombardment of Germany used an average of 634,000 gallons of 100-octane gasoline [15].

To meet this challange, President Roosevelt appointed Secretary of the Interior Harold L. Ickes to the additional position of Petroleum Administrator for War, and he, together with his Deputy Administrator Ralph K. Davies (the ranking vice-president of Standard Oil of California) forged a highly successful, if only temporary, government-industry relationship. Under this joint leadership, the petroleum industry produced 5.8 million barrels of crude oil, and over 13 trillion cubic feet of natural gas from January 1942 to August 1945. The production of gasoline increased from 462,852,000 gallons in 1941 to an astonishing 4,832,722,800 gallons in 1945 [16]. The sudden critical importance of petroleum, and its clear link to military victory, produced the two separate wartime events which would lead directly to the Louisiana, Missouri Synthetic Fuel Project of 1949.

The first of these events was the initial serious attempt to meet the huge demand for petroleum with synthetic fuel made from coal. Interestingly, it was not the military leaders, but the civilian visionaries in the government who first suggested, and then became the moving force behind, the creation of a synthetic liquid fuels program. In January 1943, Michael Straus, Director of the War Resources Council, wrote a memo to his superior, Secretary Ickes, to tell him how tired he was of "piddling around" with the tiny appropriations provided by Congress for the coal conversion research that had been carried on by the Bureau of Mines since 1935. The Bureau's scientists had been successful, after all, in producing gasoline, kerosene, and heating oil from American coals and were eager to move on to the pilot-demonstration plant stage. Yet despite these steady though unheralded successes, the Bureau's request for a modest $100,000 to design such a plant had been unceremoniously rejected by the defense-strapped Bureau of the Budget some months earlier in 1942. Consequently, Straus, supported by the director of the Bureau of Mines, R. R. Sayers, and the assistant secretary of the Interior, Oscar Chapman, appealed to Ickes for help [17]. The crusty Harold Ickes, already convinced about the future of synthetic liquid fuels, needed little further prompting. Attracted by Straus's seductive references to "your splendid record for public interest and foresight," and long deeply-suspicious of the oil companies and their lack of interest in anything which did not immediately increase their short-run profits, Secretary Ickes instructed Straus to go ahead "and see what support can be built up." [18]

After taking great pains to assure the equally-suspicious oil industry that the government did not "intend to take up the natural petroleum trade," [19] Straus and Ickes picked important Congressmen from the states with the largest coal reserves--and thus with the most to gain by the passage of a major coal-to-oil appropriations bill--to introduce and shepherd the legislation through Congress. The opening salvo was fired by the indomitable Senator Joseph O'Mahoney from Wyoming (who also happened to chair the Subcommittee on Public Lands and Surveys). He detailed for his fellow legislators the huge military and civilian demands for petroleum, lauded the scientific benefits of coal conversion, and finally introduced a bill which would authorize the Bureau of Mines to undertake a five-year, $30-million program of research which would culminate in demonstration plants to determine the cost of producing gasoline and oil from coal [20].

He was joined in his campaign by Senator C. Wayland Brooks of Illinois, Representative Jennings Randolph of West Virginia (speaking not only for the state with the largest annual coal production, but as the Chairman of the House Mining Committee), and, finally, Senator James J. Davis from Pennsylvania. By mid-summer of 1943, more than seventy witnesses--engineers, politicians, and experts of all varieties--testified before O'Mahoney's subcommittee to enthusiastically endorse the synthetic fuels program. After soothing the ruffled feathers of isolated Congressmen who saw in the enactment of the bill a further erosion of American free enterprise, Congress approved the project. The Synthetic Liquid Fuels Act became Public Law 290 on April 5, 1944.

The Act contained a major problem, the full extent of which would not become clear until the program's maturity. The project was required to be overseen by a Technical Advisory Group which met periodically to review and authorize the Bureau's projects. The problem was not in the scrutiny by an outside group, but that no less than thirteen of its eighteen members were from the nation's largest oil companies. Thus, the government's single major effort to develop a synthetic liquid fuels program was watched over by the very industry which stood to lose the most if the program became too successful. Indeed, several of these representatives--C.C. Kemp of Texas Co., E.V. Murphree of Standard Oil Development Company, and Eugene Ayers of Gulf--later held important positions on the National Petroleum Council, which ultimately discredited the Bureau program in 1952. In 1945, however, the impact of this built-in conflict was still years in the future.

The Bureau was understandably jubilant to have its many decades of service and predictions recognized at last. It set about the tasks of taking bids for the construction of several new research facilities, and trying to locate skilled technicians and other personnel in a labor-starved wartime market. These tasks, it turned out, were complicated by the very low priority rating (AA-3, allotment symbol F-6) assigned to the construction of the Bureau's research and development facilities by the War Production Board. Nonetheless, work moved slowly forward. On January 8, 1945, the Bureau signed contracts with Prack & Prack, Architects, for the design of new or expanded plants at Bruceton, Pennsylvania (laboratory work on coal conversion); Rifle, Colorado (oil shale mining and retorting); and Louisiana, Missouri (coal hydrogenation and gas synthesis) [21]. At that moment, the second major event which influenced the direction of the Missouri facility was taking place.

In the autumn of 1943, a representative of the petroleum industry suggested to the government that perhaps both parties could profit from a thorough investigation of Germany's industrial secrets. The Petroleum Administration of War was impressed with the possibilities. Since Germany was well known to have made significant advances in the field of synthetic fuels, an investigation of her secrets would allow the U. S. to take up where Berlin left off, rather than duplicating work already done. In any case, it was traditionally the right of the victor to exploit the vanquished as it saw fit. The war was still raging, however, and the idea was temporarily shelved.

More than a year passed, and on November 15, 1944, the Joint Chiefs of Staff finally authorized the creation of numerous teams, among them a Technical Oil Mission, to investigate the hidden areas of German industry. On January 8, 1945, some thirty volunteers from the Bureau of Mines and various American petroleum companies gathered in New York where they divided themselves into specialist groups to digest more efficiently the vast amount of enemy industrial data. They were made "temporary" colonels in the U. S. Army, and given a no-nonsense lecture about the urgency of their mission and the need for absolute secrecy. Thus prepared, the T.O.M. members--each an acknowledged expert in his field, well-versed in German, familiar with many of their German colleagues, and thoroughly briefed on all military-related matters--departed to join the invading Allied troops [22].

Travelling first by plane from London to Brussels or Paris, and from there by automobile to the German installations, the T.O.M. team, under the direction of W.C. Schroeder, the youthful chief of the Bureau's Synthetic Liquid Fuels Division, visited and analyzed the workings of no less than forty of Germany's largest synthetic fuel installations [23]. As each plant came under investigation, the team loaded aboard trains tons of industrial records to be shipped out of Germany to London, and eventually to Washington. At the main I.G. Farben installation in Frankfurt alone, the T.O.M. members, using work gangs of German P.O.W.s, assembled "more than a hundred tons of records...over an area larger than a city block." [24]

Of equal importance to the rescue and investigation of Germany's fuel records, was the Technical Oil Mission's success in locating and interrogating the most influential German scientists in the field of synthetic fuels. Dr. Matthias Pier, for example, the aging but brilliant father of modern hydrogenation not only responded to their skillful questioning, but ultimately instructed his very sizable staff at the Leuna plant to help the Americans as well. At the Kaiser Wilhelm Institute, the largest government-financed research laboratory in Germany the members of the T.O.M. found director Dr. Helmut Pichler and his entire staff anxious to cooperate. The list of more than 90 other influential German fuel scientists who were interrogated at length by the team members--many of whom were old friends from prewar years--reads like a "Who's Who" of German science.

Representatives of American industry could hardly wait to delve into the secrets of Germany's synthetic fuel production. Never before in the history of the modern world did a sophisticated industrial nation have at its complete disposal the industrial secrets of another nation. Among the many areas of interest, the most valuable subjects concerned the production of synthetic acetylene, alcohol, and ammonia; various catalysts; fatty acids; several processes for butadiene and synthetic rubber; fractionation; gasification; high-pressure hydrogenation; hydroforming; various patents, analytical, and testing methods [25]. "Our discoveries in Germany are of immense value in terms of national security," declared Edward B. Peck of Standard Oil Development Company. "We shall not have to go down already beaten paths in our own research and experimentation. We shall be able to profit by German mistakes and we can eliminate years of work.... [Moreover], in case of another war, development of an adequate synthetic industry would obviate...the necessity of dependence on foreign oil supplies." [26] Yet, despite this soaring optimism by both industry and government, it quickly became clear that with the end of the war and the reappearance of abundant domestic and imported petroleum, America's interest in expensive synthetic fuel began to wane sharply.

Post-War Developments

Industry was particularly reluctant to plunge ahead since even the Bureau of Mines conceded that "preliminary estimates indicate that to produce 2 million barrels of [synthetic] oil a day--less than 40 percent of our current daily consumption--would require about 16 million tons of steel and the expenditure of about 9 billion dollars.... It is doubtful that private industry will find the investment sufficiently attractive to warrant starting plant construction without governmental assistance in some form." [27] Despite the Bureau's urging that just such an expanded national synthetic oil program be undertaken, industry predictably concluded that the German experience was not economically feasible for the United States. The Department of Commerce, however, was quick to point out that "the inclination and ability of our major oil companies to commence production of synthetic fuel, should conditions require it, is well known." [28]

Whether or not the Department of Commerce realized it, conditions already required the serious support of synthetic fuels by the oil industry. The expected abundance of natural petroleum had not yet materialized. Post-war demand for gasoline and other petroleum products rose at a rate above all expectations, and by 1947, there was still no leveling-off in sight. In fact, the predicted U.S. consumption of oil products for 1948 was to exceed by 14.2 percent the wartime peak in 1945 [29]. The country was facing an energy crisis in the not-too-distant future, and the government's interest again focused on synthetic liquid fuels. This time, instead of Straus and Ickes, the movement in Congress was led by Julius Krug, the new Secretary of the Interior; the indomitable Senator O'Mahoney; various Bureau of Mines officials; Secretaries of War and of Defense, Robert Patterson and James V. Forrestal, and President Truman himself.

The oil industry became seriously concerned when such officials as Forrestal and Krug called for an $8 billion synthetic fuels program; the industry quickly lent its support to the far less threatening legislation proposed by Senator O'Mahoney that simply extended the original Synthetic Liquid Fuels Act of 1944 for three additional years. Their concern changed to anxiety, however, when Representative James Wolverton (N.J.), hoping to promote immediate action in the long-range development of synthetic fuels, introduced his H.R. 5475. Wolverton's bill would authorize the Reconstruction Finance Company to lend $400 million to anyone willing to build a coal liquefaction or shale-oil plant which could produce 10,000 barrels per day. And anxiety changed to dedicated opposition when industry realized that if no private company accepted the subsidy, the government might well build and operate the plants itself. Although the bill eventually languished to death as the result of a combination of political squabbling and the reappearance of large amounts of imported Middle East oil, the oil industry believed that its narrow escape might only be temporary. The oil men's continued political mobilization, and the general focus of their antagonism on the government's synthetic fuels research, would ultimately decide the fate of the whole Bureau of Mines program.

The program, by the way, was proceeding quite well. Since the end of 1945, the Bureau had been hard at work on the Synthetic Liquid Fuels Research and Development Laboratories at Bruceton, near Pittsburgh, to develop and improve processes for converting coal to oil. At Morgantown, on the campus of the University of West Virginia, a small laboratory was established to investigate and develop methods for producing low-cost synthesis gas--carbon monoxide and hydrogen--from coal. Members of its staff, in cooperation with the Alabama Power Company, also conducted underground coal gasification experiments at Gorgas, Alabama [30]. At Rifle, Colorado, on the United States Naval Oil-Shale Reserves, an oil-shale demonstration mine and plant was established; and at Laramie, on the campus of the University of Wyoming, another experiment station was commissioned to investigate fundamental research on the chemistry of shale-oil and on the development of more efficient retorting and refining processes. Finally, at the Northern Regional Research Laboratory in Peoria, Illinois, with funds transferred from the Bureau of Mines, the Department of Agriculture investigated the possibilities of producing synthetic liquid fuels from such agricultural residues as corncobs and the hulls of cottonseed, oats, and rice [31]. Of all these programs, the shale-oil projects looked especially promising, particularly at the 500 gallon per day plant at Rifle, Colorado. This plant produced gasoline from shale-oil at a cost of between 7.5 and 9.5 cents per gallon (or $3.15-3.99 bbl.)--still higher than gasoline from natural petroleum (3.9-4.7 cents per gallon; or $1.64-2.00 bbl.)--but certainly encouraging [32]. Yet a sizable group in the Bureau's Synthetic Liquid Fuels Division looked toward a still better prospect: coal conversion based on the technology from Nazi Germany.

The Louisiana, Missouri Project

Most of the SLF Division Engineers either had been part of the Technical Oil Mission or had aided in analyzing the detailed documents on German coal hydrogenation and gas synthesis--the flow diagrams, plant layouts, and production records--and they were impressed with the German technology. Consequently, the Bureau decided even before the end of the war to design a major coal conversion demonstration plant to test and, where possible, to improve upon the advances made by German technology. The site chosen for this Demonstration Plant was the sleepy little community of Louisiana, Missouri--about a hundred miles up the Mississippi River from St. Louis--where the government had maintained a wartime ammonia plant, the Missouri Ordnance Works. Indeed, this previous function led the Bureau of Mines to settle on the Louisiana site, since the facilities used in ammonia production could be easily adapted to the hydrogenation of coal, thus reducing the cost and time required for the opening of an installation. In mid-December 1945, the Bureau acquired the plant from the Quartermaster Corps, and the most successful and controversial demonstration plant of the early postwar years was underway.

The project ran into two immediate difficulties. The first was a problem faced by every demonstration plant involving the new areas of synthetic fuel--the inability to locate skilled workers and technicians. At Louisiana, the personnel shortage woul be solved in an unusual way. Since the demonstration plant was based on German technology, it seemed only reasonable to use German scientists. Although discussions within the Bureau considered 'importing' as many as 300 German fuel technicians, the Bureau ultimately decided to invite the seven most important German synthetic fuel experts to join the Louisiana endeavor. These scientists--Drs. Ernst Donath, Leonard Alberts Ernst Graf, Kurt Bretschneider, Hans Schappert, Walter Oppelt, and Erich Frese, all eminent men who had cooperated closely with their American colleagues in the Technical Oil Mission--jumped at the opportunity [33]. Their primary responsibility, as explained to the assembled American engineers by the plant director Lestor Hirst, was that "these people have run such plants and have the know-how. We're not going to change anything, except to add two objectives: one, we're going to apply American chemical, engineering, and instrumentation knowledge to the plant; and two, we are going to educate American manufacturers to build this kind of equipment. And that's it. Everything else is going to be just exactly the way the Germans tell us." [34]

Aside from the German scientists, however, American operators familiar with processes employing pressures and temperatures required for coal hydrogenation were simply not available. On-the-job plant training was impossible. Nonetheless, the Bureau successfully trained a total of 425 operators and technicians and prepared a host of training, operating, and safety manuals for current and future use.

The second problem concerned the plant's actual construction. After soliciting bids in March 1946, the government selected the San Francisco based Bechtel Corporation. As the contractor's engineers began to arrive in Louisiana in May, however, the Civilian Production Administration temporarily shelved the Bureau's application for authority to start construction. The Army had identified a shortage of fertilizer critical to the nation's security, and requisitioned the plant for the production of ammonia.

For the next several months, the Bureau's engineering, design, and accounting personnel went to California where they could finish their work near the Bechtel facilities. Soon after, the Bureau again took possession of the plant and pressed on to completion. Three years later, on May 8, 1949, the Bureau of Mines dedicated its nearly completed facility. To mark the occasion, the special Burlington diesel train which brought the more than 500 guests from St. Louis was powered by 1,200 gallons of synthetic oil produced from North Dakota lignite in the plant's pilot run several weeks earlier. It was the first time in this country that substantial quantities of oil had been made from coal. The principal speakers, Secretary of the Interior J.A. Krug; James Boyd, Director of the Bureau of Mines; Arno C. Fieldner, Chief of the Bureau's Fuels and Explosives Division; L.L. Hirst, Chief of the Demonstration Plant; Charles F. Kettering of General Motors; and W.C. Schroeder, Chief of the Bureau's Synthetic Liquid Fuel Division, all heralded the new demonstration plant as the door to a new and glorious age [35]. So began the bittersweet story of the Louisiana, Missouri Demonstration Plant.

The installation was not a small operation, despite the fact that its output of synthetic gasoline ranged from only 200 to 300 barrels per day depending on the type of coal and catalyst used. In construction costs alone, the plant required a substantial $10 million above and beyond the existing $17 million Missouri Ordinance Works ammonia plant which the new installation had absorbed. In addition to the more than 900 tons of structural steel used to build the demonstration plant, it used some 1,200 tons of steel tanks and pressure vessels, 40 miles of pipe and tubing, and 3 miles of sewers (see Figure 1). It was a remarkable installation from the beginning.

Much of the construction and operational information came from German plants, though it was clear that merely duplicating the German experience would not be of much value. The Germans were successful in the production of synthetic fuel, it was true, but the achievement had been under pressure of war demands (even before the war), with little thought to costs, expenditures, manpower, or the ability to close down for periodic modernization of the facilities. Thus, automatic pressure, temperature, and flow control devices had to replace hand controls wherever possible, both for higher accuracy and efficiency, and for lower production costs. The second major reason for not merely duplicating a German plant was the fact that American coal was the raw material, not German coal, and the equipment had to reflect that substantial difference. Moreover, Germany's inability to introduce modern developments during the demanding war years now forced the technicians at Louisiana, Missouri to design (and often manufacture in the plant's machine shops) daring innovations in special-alloy steel vessels, pipes, valves, instruments, pumps, and compressors--capable of withstanding the 10,300 pounds per square inch pressures and 900°F temperatures of the coal conversion processes.

Although the plant was clearly tailored to American needs, German equipment was regularly adapted for use where possible. In particular, the Bureau brought from Germany such massive and unique items as a high-pressure, forged, low-chrome steel, liquid phase converter (manufactured by Dormund-Hoerder Huettenverein), weighing 170 tons; a specialized, high-pressure, Wickel-type heat exchanger, some 54 feet long (Krupp); high-pressure hydraulic paste injection pumps; as well as a variety of high-pressure paste preheaters, lens rings, fittings, and special valves and instruments. Of particular significance was a huge Linde-Frankl Oxygen Plant, of one ton per hour capacity, at 98 per cent purity, shipped from Hochst, Germany to Missouri [36]. One of the Bureau engineers, James McGee, still shudders when he recalls the tasks of uncrating and reassembling the acres of machinery:

> First of all, the oxygen plant had been dismantled over there in Germany by GIs who knew little about such machinery; they just took

Figure 1. A view of the distillation unit at the hydrogenation plant at Louisiana, Missouri. Photograph courtesy of the United States Bureau of Mines.

cutting torches and cut wherever they thought they ought to. Anyway, these giant crates were shipped over, and stored at Louisiana, Missouri. I spent a lot of time sorting it out, and with my little German-English dictionary, tried to figure out what went where. Ultimately, we turned it over to Bechtel, since they were the architect-engineers of the plant, and told them to put it together.

...[It turned out that] some of the process piping was not shipped to this country. Some of the compressor's inner-coolers were found out worn. The compressors had to be taken apart and reconditioned. Equipment and process piping had to be cleaned and painted.... They also replaced the 50 cycle German compressor with an American 60 cycle compressor.... [37]

One of the most significant accomplishments of the Demonstration Plant was the continuous improvement of the already advanced German equipment and techniques.

The Louisiana plant, it must be realized, was actually two separate installations. The main operation was the Coal Hydrogenation Demonstration Plant, based on the Bergius-I.G. Farben process. That process was divided into two major steps: the liquid-phase hydrogenation, which liquefied the coal; and the vapor-phase hydrogenation, which converted the liquefied coal to gasoline and its byproducts. Technically (see Figure 2), the raw coal was first crushed, dried, and pulverized in a preparation area and then transferred to a 60-ton bin in the pasting area. The pulverized coal was then mixed with a tiny quantity of catalyst, such as iron oxide or tin oxalate, and with heavy oil previously obtained from the liquid-phase process, to form an oily paste. The heavy, viscous paste was put under pressure and injected by steam-driven, plunger pumps into a gas-fired tubular preheater, where it became as liquid as water as it approached the reaction temperature.

Then the paste reached the heart of the hydrogenation plant--giant liquid-phase converters, encased in protective, reinforced concrete stalls. In these massive cylinders of low-chrome steel, each weighing about 100 tons, hydrogen was reacted with the liquid coal to produce oil. The primary difference between coal and crude petroleum is that there is approximately twice as much hydrogen in the latter. Thus, for the hydrogenation of coal to gasoline, hydrogen amounting

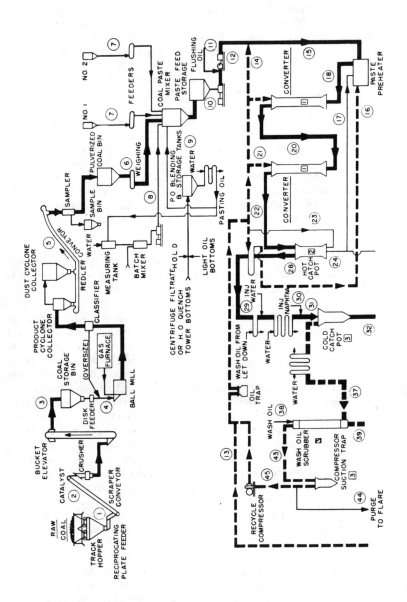

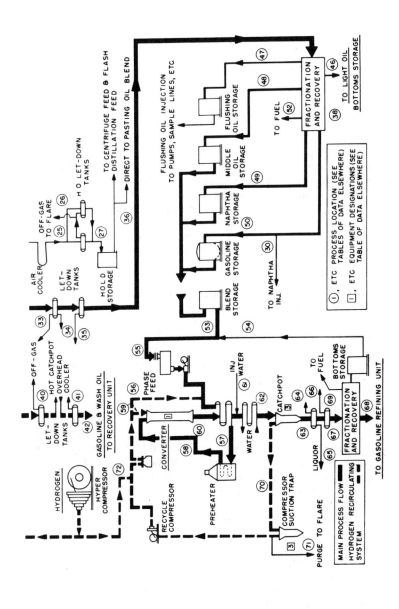

Figure 2. Process flow of coal hydrogenation plant. Source: C. C. Chaffee and L. L. Hirst, "Liquid Fuel from Coal: Progress Report on Hydrogenation Demonstration Plant," *Industrial and Engineering Chemistry*, 45, 823 (1953). Reprinted by permission.

to approximately 9 per cent of the weight of the moisture-and-ash-free coal has to be combined into the product. After about an hour in this atmosphere of high pressure and temperature (900°F), 95 per cent of the coal was converted to gas and liquid which could be used immediately or processed further. In the case of the latter, the reacted product passed through a series of hot and cold catchpots to separate the heavy oil, light oil, and gases. The heavy oil was then cleansed of ash, catalysts, and unreacted coal by centrifuging or by flash distillation. Lighter oils were also depressurized, then distilled and passed on to the vapor-phase section of the plant for further hydrogenation.

In the vapor phase, the incoming oils, cleansed and distilled, were again put under about 10,300 p.s.i. pressure, mixed with hydrogen, and forced into a heat exchanger and a preheater. The resulting vapors went into another converter where they were then passed over a fairly rugged catalyst--metal trays of fuller's earth treated with hydrofluoric acid and compounds of molybdenum, chromium, and sulfur. With each pass through this converter, about 50 per cent of the vapor was changed into gasoline [38].

The second installation at Louisiana was the Gas-Synthesis Demonstration Plant. Designed and built by the Koppers Company of Pittsburgh, the $5 million gasification unit was capable of producing between 80 and 100 barrels per day. It was completed in March of 1949, and made its first exploratory run in May. As with the hydrogenation demonstration plant, the many complex problems taxed even the formidable ingenuity of the Bureau's and contractor's engineers though, like the hydrogenation unit, much of the original German data had been improved on at the pilot plants of Bruceton. As a result of these modifications, and others from the commercial, fluidized catalyst bed method developed by the Carthage Hydrocol Company in Brownsville, Texas, the Bureau's gas-synthesis process bore little resemblance to the original Fischer-Tropsch process.

Briefly (see Figure 3), in the Gas-Synthesis Demonstration Plant the incoming coal was first crushed, pulverized to 200-mesh, and then dried. The powdered coal was then suspended in oxygen and, accompanied by superheated steam, fed into a gasifier--a refractory-lined steel shell, some 6 1/2 by 9 feet in inside measurement. There, the conversion took place at more than 2,000°F. This particular stage was awesome, since the gasifier required about 28 tons of coal, 24 tons of oxygen, and 35 tons of superheated steam daily, in order to produce 2 million cubic feet of raw synthesis gas. The large amounts of oxygen required for the reaction, incidentally, came from the Linde-Frankl unit which had been brought from Germany; operating at 300°F below zero, it extracted oxygen from air [39].

The next stage--the purification of the synthesis gas--was no less complex. Once cooled, the gas entered a series of precipitators and separators, catalytic converters, iron oxide boxes, and activated carbon absorbers, to filter out the minute quanitites of sulfur, tar dust, resins, hydrogen cyanide, and so forth. The resulting purified synthesis gas was then converted to a liquid in another reactor, at 450 to 550°F., and over a "jiggling-bed" catalyst developed by the Bureau of Mines at Bruceton. The mixture of gases was cooled to condense as much a possible of the liquid products, which were then passed to the final distillation and refining stages. By this process, the Gas-Synthesis Demonstration Plant at Louisiana produced daily about 55 barrels of finished gasoline, 10 barrels of diesel oil, 12 barrels of heavy oils and waxes, and 5-10 barrels of propane [40]. Through cracking, the wax could be converted to diesel fuel and lubricants, or used directly in polishes, insulating materials, and the chemical industry.

For the next three years, from the end of 1949 through 1952, the two installations produced a variety of fuels and byproducts. Technical innovations abounded and, by all accounts, the Louisiana demonstration plants hummed with the activity and cordiality of dedicated experts whose time had finally come. There were difficulties aplenty, with injector pump, instrument, and converter coke problems. For example, at the beginning of a particular vapor-phase run, conducted at 10,000 p.s.i., a cross leak in the feed-product exchanger spoiled the entire product and necessitated a complete shutdown. But the Bureau engineers and the German scientists managed to solve each problem as it occurred. Moreover, they made significant improvements in the design of high-pressure injection pumps, in the development of various catalysts, in the increased efficiency of gas synthesis, and in the creation of an entirely new feed system for the coal gasification process [41], to name but a few.

An indication of the level of advances made by the engineers at Louisiana was the substantial improvement over the German plants which American scientists originally so admired. For example, the thermal efficiency of the average German wartime plant, calculated at 28.9 percent, was improved by American engineering to an impressive 55 percent [42]. The Louisiana plants also made extensive tests on various types of American coal, converting 4,000 tons of northern Wyoming sub-bituminous coal from the Lake DeSmet area

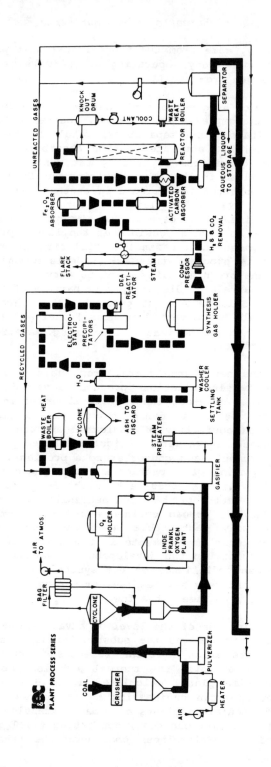

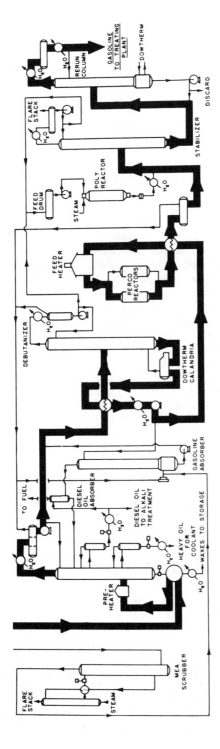

Figure 3. Flow sheet of Bureau of Mines modification of gas synthesis process. Source: L. L. Hirst and R. G. Dressler, "An American Fischer-Tropsch Plant," *Industrial and Engineering Chemistry*, 44, 458 (1952). Reprinted by permission.

into 273,000 gallons of fuel in one six-week period. They then moved on to the conversion of Pittsburgh-seam coal from Washington County, Pennsylvania, during the next period; then Illinois No. 6 coal in the next period; then western Kentucky coal; and North Dakota coal after that. The detailed reports and publications which resulted are models of astute scientific research [43]. The Louisiana engineers, in compliance with the Synthetic Liquid Fuels Act's instruction to furnish industry with basic information and cost data for the development of a synthetic fuels industry, produced a remarkable 83-page, mimeographed report, "Estimated Plant and Operating Costs for Producing Gasoline by Coal Hydrogenation," which outlines, in mind-numbing detail, every imaginable cost in the construction and maintenance of a 30,000 barrel per day installation [44]. Though numerous problems continued to plague the Bureau of Mines installation--generally in the areas of catalysts, and high-pressure vessels--the staff of the Nation's first coal-to-oil demonstration plants had every reason to be pleased. Moreover, as the result of Senator O'Mahoney's Synthetic Liquid Fuels Extension Act, the government made additional funds available on July 2, 1948 to conduct research on refining problems, and it was clear that more money was to follow. The rising tide of Cold War hysteria focused public attention on America's national security, one key to which was a domestic source of petroleum. The development of synthetic liquid fuels relegated for so long to the backwater of industrial progress was suddenly catapulted to the highest of priorities.

By mid-1949, with the Cold War rapidly heating up and references to a Third World War with the Soviet Union becoming commonplace, Congressmen of every political stripe spoke eloquently on the importance of an uninterruptable supply of oil for national defense. Many officials pointed out the increasing civilian requirements; the nation's total reliance on gasoline-driven transportation and the critical need for the many chemical byproducts produced from petroleum. The final argument, however, was simply that "should World War III come, ...with much of the naval shipbuilding activities of Russia reported to be directed toward modernized submarines, it becomes obvious even to the layman that overseas tanker lines from the Near East and South America would be difficult to maintain." [45] Consequently, on June 16, 1949, both the National Security Resources Board and the State Department reported favorably on Wolverton's bill (H.R. 5475) which called for massive RFC subsidies to help in the construction of new synthetic fuel plants. To ensure that the management of such a program would be in the most responsible hands, President Truman issued Executive Order 10161 placing the development of domestic petroleum sources, including synthetic fuels and associated hydrocarbons, in the hands of the Secretary of the Interior, and thus the Bureau of Mines. The oil industry, already alarmed at the government's growing interest in synthetic fuels, now prepared for a fight.

Industry apprehension increased substantially when, the following year, in 1950, the fear of national unpreparedness in the face of a looming conflict with the communist world led to the passage of the Defense Production Act. High on the list of national defense items to be developed quickly was liquid fuel. The Bureau of Mines answered the call with a remarkable report entitled, "Justification for Construction of Coal Hydrogenation Plants," a follow-up to its 1949 report on the estimated costs of a 30,000 barrel per day installation. After forty pages of closely-analyzed economic, technical, and cost-efficiency data, the Bureau concluded that:

1. There is ample justification for the construction of two 15,000 barrel-per-day coal hydrogenation plants, costing a total of $326 million, as a defense step.

2. In view of the importance of oil to national defense and the growing dependence on imports (plus danger of losing Middle East oil), it is imperative that the two plants be constructed immediately.

3. It is entirely feasible to begin construction of commercial coal hydrogenation plants without delay. The process has been used extensively on a large scale and the two proposed plants involve no untried processes....

5. These plants, in additon to paving the way for the pending large-scale synthetic fuel industry, will provide important quantities of critical materials now in short supply, including 34 million gallons annually of benzene, and 125 million gallons of aromatics for aviation gasoline and explosives....

7. There are competent and responsible firms that are willing to build and operate the proposed plants....

8. All funds for construction of the plants can be borrowed from private sources, with

purchase contracts and loan guarantees as provided for in the Defense Production Act of 1950.

9. On the basis of cost and marketing studies, the total govenment appropriations for this project and the ultimate cost to the government will be considerably less than the construction costs of the plants. On the basis of present conditions it is estimated that the investment can be recovered with 3 percent interest in 11.5 years. [46]

Here was not only the first step toward a national synthetic petroleum program but, since the Bureau drew largely for its findings on the Louisiana demonstration plants, a glowing evaluation of that installation's success and importance. The Missouri plant, it seemed, was the standard-bearer in America's march toward fuel self-sufficiency.

Mobilization of Opposition

The next two years saw little to diminish the Bureau's confidence in the Louisiana, Missouri plant. The annual Bureau reports to Congress on its activities of the past year continued to note in evaluating the performance of each experimental or pilot plant that Louisiana produced "good results," and that "early reports are satisfactory." [47] The annual report in 1951 remarked that "each [coal-to-oil] run was successful and achieved its objectives." [48] In its 1952 report, the Bureau once again cited the Louisiana installations as having "provided added data for the design and operation of prototype commercial synthetic fuel plants." [49] Yet, beneath the surface of what appeared to be routine approval of the installations, there was a storm brewing. The Louisiana demonstration plants would not survive that storm.

First, there was a small but growing number of both industry and Bureau personnel who simply did not believe that the information gleaned from the operation was worth the heavy allocation of funds and manpower. Modified German data was interesting, to be sure, but not necessarily valuable. For example, E.L. Clark of the Bureau's Synthetic Liquid Fuels Division in Washington reduced the findings of the installation to the following impressions:

...*first*, a sense of admiration for the tremendous accomplishments. . .achieved by German chemists and engineers; second, a realization that less than 60 percent of the coal is converted to the desired distillable oil; third, the stoichiometry shows that only about half of the hydrogen consumed is found in this distillable product oil; fourth, approximately 10 percent of the oil produced is lost in the solids-purging operations; and, finally, the problems of handling mixtures of heavy oil and solids are different and costly [50].

The second group opposed to the Louisiana plant came entirely from within the Bureau and revolved around the influential H.H. Storch--soon to become the Chief of the Synthetic Liquid Fuels Division. The former director of the Brucetown experiment station, Storch believed that the pace of the demonstration plants had run too far ahead of the Bureau's experimental knowledge, and he called for a return to fundamental laboratory and pilot-plant research. Broadly speaking, Storch believed that the main objectives of both the coal-hydrogenation and gas-synthesis processes had been achieved by 1952: "Several representative American coals of various ranks had been converted into liquid fuels, using American techniques and equipment and found satisfactory... [and] the processes demonstrated were technically feasible." What was needed now, however, was "several years of laboratory and pilot-plant research." [51]

The third group which opposed the continuing operation of the Louisiana installations--generally representatives of the oil industry--pointed to the issue of cost per gallon. Indeed, the question of cost-efficiency would become the deciding feature in the debate over the demonstration plant's future. Several points must first be made before examining the various sides of such an argument. The first is that during the initial years of industrial transition from a natural product to its synthetic form, the synthetic is always more expensive--whatever the item. Second, the development of a synthetic fuels program in the United States was essentially a World War II and Cold War creation which, like the development of the German program, was created to produce a stable source of critical petroleum products without any real concern for its cost per gallon justification. Moreover, the debate over the economics of synthetic fuel in the United States was based on the fluctuating production figures of test installations whose capacity never exceeded 300 barrels per day. Lastly, the cost figures varied widely, depending on the type of coal tested and the changing value of the by-products.

The Bureau of Mines had never implied that the production of synthetic liquid fuels was competitive with natural petroleum. In both of its comprehensive reports on the cost of coal conversion—the 1949 report estimating the cost of a 30,000 barrel per day plant, and the 1950 report justifying the urgent need for two 15,000 barrel per day plants as the start of a larger program [52]—the Bureau clearly acknowledged the initially higher cost of synthetic petrochemicals. The 1950 report, in particular, was written with the knowledge that its findings would be used as a political football. Consequently, it opened with an almost defiant reminder that "the Defense Production Act does not require that projects be economically justified—in fact, provision is specifically made for defense projects that are uneconomic." [53] Having thus defanged its critics, the report launched into a detailed examination of every facet of two hydrogenation plants at assumed locations in Wyoming and Kentucky, ending with the estimate that the production cost of gasoline—before profit or taxes, and defrayed by the sale of its byproducts ammonium sulfate and sulfur—was 10.2 cents per gallon [54]. By the time it reached the gas station pump, it could be 15 or 16 cents per gallon, about 25 percent higher than the current cost of natural petroleum gasoline.

The oil industry found little difficulty in turning these figures to its advantage. The Bureau realized its predicament, of course, but could do little to avert the collision. Indeed, having stated all the facts, the Bureau of Mines could now only place the issue before the petroleum industry that it was charged by the Synthetic Liquid Fuels Act of 1944 to serve. On April 21, 1950, the new Secretary of the Interior, Oscar Chapman, commissioned the industry's National Petroleum Council (1) to review the data and estimated costs of producing synthetic liquid fuels; (2) to prepare an independent cost estimate; and (3) to recommend any means for the improvement of future cost estimates by the Bureau of Mines.

The Council turned over the project to an NPC Committee on Synthetic Liquid Fuels Cost, composed of representatives of the nation's 47 largest petroleum and coal companies, virtually all of whom had at one time or another spoken out against synthetic fuels. For a year and a half the Committee labored, drawing on more than 100 other companies for information, and employing an additional 105 technicians. The Interim and Final Reports are measured better in pounds than in pages. Industry's estimate of synthetic fuel cost, as expected, was substantially higher than the Bureau's estimate of 10.2 cents per gallon. The National Petroleum Council's Committee of Synthetic Liquid Fuels Cost reevaluated every factor from the cost of housing for the workers, to requirements of steel and utilities, and determined the cost of gasoline by coal hydrogenation at an average of 43.5 cents per gallon! [55]. Moreover, the NPC report was severely critical of the Bureau's methods and calculations. The Bureau's Synthetic Liquid Fuels Division was in trouble and knew it.

Frustrated in its efforts to refute the NPC cost estimates and criticism, the Bureau of Mines, led by Secretary Chapman, engaged Ebasco Services, Inc., a prominent engineering, management, and consulting firm, to make an independent review of all the facts, again including housing cost, manpower, and the fluctuating market value of the salable byproducts. They were not shown the National Petroleum Council's report. This time it was Ebasco's turn to labor, for six months, until the publication of its findings in March 1952. Their findings agreed with neither the Bureau of Mines' nor the National Petroleum Council's--though they were certainly closer to the former. In fact, Ebasco only raised the Bureau's esimate from 10.2 cents per gallon to 11.4 cents per gallon! While the price was still prohibitively high (acknowledged by Ebasco's own declaration that "We do not believe it would be feasible to finance the projects...with private capital....," [56]), the Bureau's Synthetic Liquid Fuels Division believed its work had been vindicated, and that, regardless of costs, the nation ought to embrace a synthetic fuels program as an investment in national security. The National Petroleum Council, meanwhile, issued a counter-report defending its procedures and findings, and pointed out a variety of errors in the Ebasco report which would raise the estimate to 28.1 cents per gallon [57].

Through the summer and fall of 1952, all 3 parties--the Bureau, the NPC, and Ebasco--issued additional reports, interim reports, and counter-reports [58]. Each reevaluation of the estimated future value of such byproducts as benzenes, phenols, and creosols; of housing costs for refinery workers; and of anticipataed tax, interest, and profit levels, altered the final cost per gallon. So greatly did their findings vary, ultimately ranging from 1.6 cents per gallon to 41.4 cents per gallon, that any firm conclusion was extremely difficult. For example, the National Petroleum Council compared its findings with those of the Bureau and of Ebasco as shown in Table 1. The Bureau of Mines, on the other hand, saw comparisons differently (Table 2). The figures, however, were only pawns in a larger game. It was, in reality, a deadlock between the Bureau of Mines and the oil industry. The Bureau saw itself as the unheeded messenger from the future, committed to preparing the American public for the time when natural petroleum would become too expensive or unavailable.

Table 1. Summary of Unit-Cost Estimates of Coal Hydrogenation: National Petroleum Council, U. S. Bureau of Mines, and Ebasco Services, Inc.

Cost Items	NPC (10/31/51)	Ebasco (3/52)	Bureau (10/25/51) of Mines
Manufacturing	25.3	19.5	17.7
Housing Costs	2.6	- - -	- - -
Financial	19.0	22.0	8.2
Total costs	46.9	41.5	25.9
Less byproducts revenue	-5.5	-13.4	-14.9
Gasoline costs/gal	41.4	28.1	11.0

"Source: NPC, "Interim Report of the NPC's Committee on Synthetic Liquid Fuels Production Cost," July 29, 1952, p. 8.

The oil industry saw itself defending the principle of free enterprise against government encroachment or even eventual nationalization [59], a principle made easier to defend with the appearance of unprecedented amounts of inexpensive oil from the Middle East. As if looking into the future, the Bureau of Mines already acknowledged that "for the case in question, the demonstration coal-hydrogenation plant at Louisiana, Missouri, has served as a scapegoat...." [60]

The deadlock was broken sooner than anyone had expected, only months later, in January 1953, with the Presidential inauguration of Dwight D. Eisenhower. Unlike the Truman administration which had supported the development of synthetic liquid fuels, the incoming Republicans not only shared the oil industry's concern about excessive government involvement in the private sector, but had pledged the voters to cut the rate of federal spending. The prospects for synthetic fuels further worsened when the National Petroleum Council issued its final report in February which concluded that "all methods of manufacturing synthetic liquid fuels proposed by the Bureau of Mines are definitely uneconomical under present conditions.... [Moreover], the need for a synthetic liquid fuel industry in this country is still in the distant future.... We question the wisdom of the Government financing large-scale demonstration plants [when] such techniques can be developed from well designed pilot plants at a small fraction of the cost of so-called demonstration plants." [61] The Bureau watched as its opponents began to close ranks.

The NPC's conclusion that "well-designed pilot plants" were more logical than large demonstration plants, was also a philosophy shared by H.H. Storch and his supporters in the Bureau. Whether by coincidence or design, Eisenhower's new Secretary of the Interior, Douglas McKay, appointed Dr. Storch to head the Bureau's Synthetic Liquid Fuels Division! It was the beginning of the end for the Louisiana, Missouri project.

The stunning array of opposition to the Bureau's synthetic fuels program moved so swiftly that the battle was over before the engineers at Louisiana realized what had happened. When the House Appropriations Committee opened its budget hearings in March 1953, its first official act was to cut two million dollars earmarked for the Louisiana plants from Interior's budget. After several days of consultation with Bureau of Mines officials and Interior Secretary McKay, the Committee's Chairman, Homer Budge of Idaho, delivered the <u>coup de grace</u> by cutting the Bureau of Mines' budget by $9 million, thus terminating the entire synthetic fuels program. It was all over by March 11, 1953. To add insult to injury,

Table 2. Comparison of the Estimated Costs (1952) for Coal-Hydrogenation Gasoline (Single Plant)

		Bureau of Mines Data	Ebasco Services Date	National Petroleum Council Date
Plant size	bbl./stream-day	33,000	30,000	30,000
Stream year	days	330	330	330
Total investment	dollars	414,400,000	403,827,000	437,060,000
Operating cost (including depreciation)	annual dollars	53,199,500	58,753,000	68,060,000
Production, annual:				
Gasoline	bbl.	6,773,800	6,815,800	7,113,900
L.P.G.	bbl.	2,590,480	1,590,480	2,332,350
Benzene	gal.	11,750,000	11,750,000	—
Toluene-zylene	gal.	39,350,000	39,350,000	—
Phenol	lb.	43,800,000	43,800,000	31,025,000
Cresols	lb.	71,000,000	56,600,000	50,370,000
Xylenols	lb.	25,000,000	25,000,000	66,430,000
Ammonium sulfate	tons	131,000	131,000	—
Sulfur	tons	17,000	17,000	—

Table 2. (continued)

	Bureau of Mines Data	Ebasco Services Date	National Petroleum Council Date
Cost of total liquid products........ cents/gal.	10.2	11.4	16.4
Value of byproducts ..annual dollars	48,559,000	44,692,000	16,487,000
Cost above byproducts returns ......annual dollars	4,640,500	14,061,000	51,573,000
Cost of gasoline, cents/gal.:			
Before profit and income tax	1.6	4.9	17.3
At 6% return on total investment after 50% income tax......	19.1	21.8	34.8

Source: Bureau of Mines Synthetic Liquid Fuels Program, 1944-1955, Part I: Oil From Coal, Report of Investigation 5506, Table 21, p. 210.

Representative Budge not only praised the Bureau for the "great progress [that] had been made," but, after all the NPC reports to the contrary, he justified the termination of the synthetic liquid fuels program on the premise that "they had come within a couple of cents of the commercial cost of producing gasoline from petroleum." [62] Consequently, the new Administration concluded it was time to end the government's involvement and turn the synthetic fuel program over to the private sector.

Numerous politicians, especially Democrats from coal-rich states, rose up in anger at the oil lobby's blatant scuttling of the synthetic fuels program. Their antagonism was compounded by Budge's admission that the Bureau's work in coal hydrogenation and shale-oil conversion appeared to be competitive with natural petroleum after all. "The oil companies tried for a long time...to close down this plant [at Louisiana, Missouri]," thundered Senator Estes Kefauver of Tennessee, "because they do not want the competition of oil from coal.... It now appears that the [oil] interests have prevailed." [63]

The engineers at Louisiana did not have to wait long to learn how all this activity in Washington would affect them. On April 1, L. L. Hirst, chief of the installation, was notified that both the hydrogenation and gas-synthesis operations were to be shut down in 60 days! There was no appeal or last-minute reprieve. The engineers and German scientists--some sad, most outraged--disbanded, and the operation of the plant was brought to a halt. In mid-June, scarcely three months after the hearings on the Interior Department's budget under the new Eisenhower administration, the Louisiana plant was officially closed, and turned over to the Department of the Army for disposal. The installation was soon sold to the Hercules Powder Company, still the owners today. In his annual report to Congress for 1953, Interior Secretary McKay blandly reported the closing of the coal-to-oil demonstration plants at Louisiana, citing as the reason, the "need for additional scientific and engineering information basic to future industrial development." [64]

Most of the American engineers remained with the Bureau of Mines--luminaries like L.L. Hirst, James P. McGee, Howard W. Wainwright, A.C. Fieldner, Sidney Katell. Almost all work at the synthesis-gas-from-coal pilot plant at Morgantown, West Virginia. One American who did not remain with with the Bureau was W.C. Schroeder, former Chief of the Synthetic Liquid Fuels Division and a lifelong advocate of coal-to-oil conversion. With the termination of the Louisiana plant, Schroeder saw any plans for a national synthetic fuels program recede into the distance. He quit the Bureau of Mines and spent the intervening years as a consultant to private companies, and at present is a professor of chemical engineering at the University of Maryland [65].

Epilogue

For the next twenty years after the end of the Louisiana, Missouri Synthetic Liquid Fuels Demonstration Plants, the oil industry grew more and more dependent on relatively inexpensive petroleum from the Middle East and Venezuela. Between 1953 and 1973, in fact, America's annual reliance on foreign oil rose from less than one percent (19 million barrels) to more than 29 percent (1.1 billion barrels). While the oil industry did, indeed, invest heavily in the nation's great coalfields during the same period, it was a program prompted by a diversification of its huge holdings, rather than with a view toward future synthetic fuel production. Then the Arab oil embargo and the subsequent price hikes of 1973-1974 suddenly caused the government's attention to be riveted once again on the question of synthetic liquid fuels.

Energy-related legislation poured out of Congress. The Energy Supply and Environmental Coordination Act of 1974 (Public Law 93-319), for example, initiated a thorough investigation of the nation's fuel supplies, and allowed the temporary suspension of certain air pollution restrictions regarding the burning of fuels. On October 8th, Congress passed the Energy Reorganization Act of 1974 (Public Law 93-438) to create an Energy Research and Development Administration. On December 31st, lamenting the Nation's "past and present failure to formulate a comprehensive and aggressive research program," and calling for a national commitment "similar to those undertaken in the Manhattan and Apollo projects," [66] Congress passed the Federal Nonnuclear Energy Research and Development Act of 1974. By this Act (Public Law 93-577), Congress allocated nearly half a billion dollars "to accelerate the commercial demonstration of technologies for producing syncrude and liquid petroleum products from coal." [67]

That little serious progress ensued may be deduced from President Carter's speech before a special joint session of Congress, on April 20, 1977, to announce a national energy program which would be "the moral equivalent of war." High on the list of the nation's energy goals, the President declared, would be a significant increase in the use of coal, and especially "clean energy sources like liquefied and gasified coal...[on which there has been]...very little research and development." [68] The history of the development of synthetic liquid fuels in the United States had come full circle.

References and Notes

1. Bureau of Mines, <u>Report of Investigations 4942, Part 1, Summary, 1952</u> (G.P.O., Washington, 1953), p. 1.

2. See Arnold Krammer, "Fueling the Third Reich," <u>Tech. and Cult.</u>, 19, 394 (1978).

3. U.S. Congress, House Committee on Mines and Mining, <u>Production of Gasoline from Coal and Other Products</u>, 77th Cong., 2nd sess., June 18, 1942, pp. 28-35, 57; Kenneth Gordon, "The Development of Coal Hydrogenation by Imperial Chemical Industries, Ltd.," <u>J. Inst. Fuels</u>, 9, 69 (1935) and C. Cochran and E.W. Sawyer, "Hydrogenation at Billingham in Retrospect," <u>Industrial Chemist</u>, (May, 1959), pp. 221-29.

4. Rouillé, "La Technique Réalisée a Liéven pour L'Hydrogénation des Combustibles Solides Selon les Procédés de la Société Nationale de Recherches" [Hydrogenation of Solid Fuels at Lieven According to the Processes of the Société Nationale de Recherches], <u>Mem. soc. ing. civils. France</u>, 90, 746 (1937); "Calais Petrol-From-Coal Works To Be Abandoned," <u>Chem. Age</u>, 91, 265, (February 15, 1964).

5. R.W. Rutherford, "Oil From Coal in Japan," <u>Coke and Smokeless-Fuel Age</u>, 6, 68 (1944).

6. Harold F. Williamson, et al., <u>The American Petroleum Industry: The Age of Energy, 1899-1959</u> (Northwestern University Press, Evanston, 1963), p. 748; <u>Federal Register</u>, June 7, 1941, pp. 2760.

7. Ibid.

8. Gabriel Kolko, "American Business and Germany, 1930-1941," <u>Western Political Quarterly</u>, 15, 719 (1962).

9. Arno C. Fieldner, "Frontiers of Fuel Technology," <u>Chem. and Eng. News</u>, 26, 1700 (1948).

10. V.H. Manning, "Experiment Stations of the Bureau of Mines," <u>Bureau of Mines, Bulletin 175</u> (1919).

11. Arno C. Fieldner, Henry H. Storch, and Lester L. Hirst, "Bureau of Mines Research on Hydrogenation and

Liquefaction of Coal and Lignite," Am. Inst. Min. and Met. Eng.; Tech. Pub. 1750 (August 1944), p. 1.

12. Fieldner, "Frontiers of Fuel Technology."

13. Benjamin T. Brooks, Peace, Plenty, and Petroleum (The Jacques Cattell Press, Lancaster, 1944), p. 61. Later technological advances would increase the American yields to 130-135 gallons per ton: U.S. Congress, Senate, Congressional Record, 78th Cong., 1st sess., 89, pt. 7, November 9, 1943, p. 9328.

14. Ibid.

15. U.S. Congress, Senate. Investigation of Petroleum Resources, Senate Resolution 36. 78th Cong., 1st sess., June 19-25, 1945, pp. 94, 59, 266-7, 282-3; J. Stanley Clark, The Oil Century: From the Drake Well to the Conservation Era (University of Oklahoma Press, Norman, 1958), pp. 139-40.

16. John W. Frey and H. Chandler Ide, A History of the Petroleum Administration for War, 1941-1945 (G.P.O., Washington, 1946), pp. 7, 243, and statistical tables in Appendix 12; pp. 438-57. For efforts in reducing fuel, see J.F. Barkley, Thomas C. Cheasley, K.M. Waddell, The National Fuel Efficiency Program During the War Years, 1943-1945, Bureau of Mines Bulletin 469, (1949).

17. Michael Straus to Harold L. Ickes, January 28, 1943, RG 48 (Dept. of the Interior), Central Classified Files, 1937-1953, Box 3762, File 11-34 (Synthetic Fuels), National Archives. For an outstanding analysis of these congressional machinations, see Richard H.K. Vietor, "The Synthetic Liquid Fuels Program: Energy Politics in the Truman Era," (Working Paper, Harvard Business School, HBS 78-54).

18. Harold L. Ickes to Michael Straus, January 29, 1943.

19. Michael Straus to Ralph K. Davies, February 1, 1943.

20. U.S. Congress, Senate Subcommittee on Public Lands, Hearings on Synthetic Liquid Fuels, 78th Cong., 1st sess., August 1943.

21. Report of the Secretary of the Interior on the Synthetic Liquid Fuels Act from January 1, 1945 to December 31, 1945, Mimeographed File No. 124830, pp. 15-16.

22. Albert E. Miller, "The Story of the Technical Oil Mission," pres. at 25th Annual Meeting of the Am. Petrol. Inst., Chicago, Illinois, November 14, 1945, mimeographed, p. 3.

23. British Intelligence Objectives Subcommittee (BIOS), Inspection of Hydrogenation and Fischer-Tropsch Plants in Western Germany During September, 1945, CIOS/BIOS Final Report 82 (London, November 1945).

24. Joseph E. DuBois, Jr., The Devil's Chemists (Beacon Press, Boston, 1952), p. 36.

25. W.C. Schroeder, "Capture of German Records will Boost United States Synthetic Fuel Development," Oil and Gas J., 44, 78 (June 30, 1945); W.C. Schroeder, "Technical Oil Mission Studies German Petroleum Research Activities," Oil and Gas J., 44, 113 (November 24, 1945); and W.C. Schroeder, "Manufacturing German Synthetic Liquid Fuels, Gas Age, 96, 21 (November 29, 1945); Richard L. Kenyon, "The German Chemical Industry, Past and Present," Chem. and Eng. News, 25, 1437 (1947); and "Fischer-Tropsch Held One of Most Valuable Enemy Processes," Chem. and Eng. News, 25, 839 (1947).

26. Ruth Sheldon, "The Hunt for Nazi Oil Secrets," Saturday Evening Post, October 6, 1945, p. 122.

27. Report of the Secretary of the Interior on the Synthetic Liquid Fuels Act from January 1, 1947 to December 31, 1947, mimeographed, File No. 30409, iii.

28. R.L. Trisko, "United States Petroleum Import Prospects," Industrial Reference Service, Part II: Metals and Minerals, 5, 11 (July 1947). The United Kingdom, incidently, found itself in similar economic straits, and in the Ministry of Fuel and Power's so-called "Gordon Report" of 1947, also elected to ignore the mountains of German industrial data which its own investigators had worked so hard to obtain. See British Ministry of Fuel and Power, Report on the Petroleum and Synthetic Oil Industry of Germany, BIOS Overall Report #1 (London, 1947), p. 3; Kenneth Gordon, "Progress in Hydrogenation of Coal and Tar," Chem. Age (October 28, 1946), pp. 795-804.

29. Report of the Secretary of the Interior....1947, p. i.

30. James L. Elder, M.H. Fies, Hugh G. Graham, R. C. Montgomery, L.D. Schmidt, and E.T. Wilkins, "The Second

Underground Gasification Experiment at Gorgas, Alabama," Bureau of Mines, Report of Investigations 4808, and 5367 (1951, 1957).

31. For a detailed description of each plant, see Report of the Secretary of the Interior.... 1945, pp. 1-44 passim; R.C. Grass and H.H. Storch, "Coal-to-Oil Research at Bruceton, Pa.," Chem. and Eng. News, 28, 646 (1950); Synthetic Liquid Fuels. Annual Report of the Secretary of the Interior for 1950; Bureau of Mines, Report of Investigations 4773 Part IV: Oil From Secondary Recovery and Refining (1951).

32. "Synthetic Liquid Fuels Studies from Several Viewpoints," Chem. and Eng. News, 26, 610 (1948); Trisko, Industrial Reference Service, op. cit., pp. 2, 10-11.

33. For a verbatim record of the frank discussions of the numbers and problems involved in bringing such German scientists to the United States, see The proceedings of the Technical Oil Mission, Held Under the Auspices of the Petroleum Administration for War and the Bureau of Mines, mimeographed, File No. 10160, December 13-14, 1945, pp. 89-98. In addition to the synthetic liquid fuels projects, German scientists were brought to the U.S. to participate in a variety of other ventures; see Joint Chiefs of Staff, "Statistical Report of the Aliens Brought to the United States Under the Paperclip Program, December 1, 1952," Office of Technical Services, Department of Commerce; and Clarence G. Lasby, Project Paperclip: German Scientists and the Cold War (Atheneum, New York, 1971). Relations with former enemy scientists quickly became so congenial that it became popular to regard the German scientists (or all scientists for that matter) as neutrals in the previous war. See Gavin de Beer, The Sciences Were Never At War (Nelson, London, 1960), and "Science Has No Nationality," Science Illustrated, 2, 13 (February 1947).

34. Inverview with James McGee, Morgantown, West Viriginia, August 2, 1977.

35. "Bureau of Mines Dedicates Coal-to-Oil Demonstration Plants," Chem. and Eng. News, 27, 1520 (1949).

36. Report of the Secretary of the Interior....1947, pp. 31-2.

37. James McGee, op. cit.

38. E.A. Clarke, C.C. Chaffee, and L.L. Hirst, "Early Operations of the Hydrogenation Demonstration Plant," <u>Bureau of Mines, Report of Investigations 4944</u> (1953); J.A. Markovits, "Coal Hydrogenation: Summary of Process and Brief Description of Special Equipment Used in Demonstration Plant at Louisiana, Missouri," <u>Mech. Eng.</u> (July 1949), pp. 553-59; C.C. Chaffee and L.L. Hirst, "Liquid Fuel From Coal: Progress Report on Hydrogenation Demonstration Plant," <u>Ind. Eng. Chem.</u>, 45, 822 (1953); Merritt L. Kastens, L.L. Hirst and C.C. Chaffee, "Liquid Fuel From Coal," <u>Ind. Eng. Chem.</u>, 41, 870 (1949).

39. H.R. Batchelder, R.G. Dressler, R.F. Tenney, L.C. Skinner, and L.L. Hirst, "The Role of Oxygen in the Production of Synthetic Liquid Fuels From Coal," <u>Bureau of Mines, Report of Investigations 4775</u> (1951).

40. J.A. Markovits and L.L. Hirst, "Two American Coal-to-Oil Demonstration Plants," <u>Mines Mag.</u> (Merritt L. Kastens, L.L. Hirst and C.C. Chaffee, "Liquid Fuel From Coal," <u>Ind. Eng. Chem.</u>, 41, 870 (1949).

39. H.R. Batchelder, R.G. Dressler, R.F. Tenney, L.C. Skinner, and L.L. Hirst, "The Role of Oxygen in the Production of Synthetic Liquid Fuels From Coal," <u>Bureau of Mines, Report of Investigations 4775</u> (1951).

40. J.A. Markovits and L.L. Hirst, "Two American Coal-to-Oil Demonstration Plants," <u>Mines Mag.</u> (Colorado), November 1952, pp. 129-144; M.L. Kastens, L.L. Hirst, and R.G. Dressler, "An American Fischer-Tropsch Plant," <u>Ind. Eng. Chem.</u>, 44, 450 (1952); L.P. Wenzell, Jr., R.G. Dressler, and H.R. Batchelder, "Plant Purification of Synthesis Gas," <u>Ind. Eng. Chem.</u>, 46, 858 (1954).

41. J.T. Donovan, B.H. Leonard, J.A. Markovits, "Design and Development of High-Pressure Injection Pumps for Hydrogenation Service," <u>American Society of Mechanical Engineers (ASME) Paper</u> No. 51-A-71, November 25-30, 1951; <u>Synthetic Liquid Fuels, Annual Report of the Secretary of the Interior for 1951, Part 1: Oil from Coal, Bureau of Mines, Report of Investigations 4865</u> (1952), p. 9; H.R. Batchelder and L.L. Hirst, "Coal Gasification at Louisiana, Missouri," <u>Ind. Eng. Chem.</u>, 47, 1522 (1955). Arnold Krammer, "American Adaptation of German Synthetic Fuel Technology; 1945-1953, American Society of Mechanical Engineers, Publication 80-Pet-50 (February 1980).

42. L.C. Skinner, R.G. Dressler, C.C. Chaffee, S.G. Miller, and L.L. Hirst, "Thermal Efficiency of Coal Hydrogenation," Ind. Eng. Chem., 41, 87 (1949).

43. See, for example, R.M. Busche, H.R. Batchelder, and W.P. Armstrong, "A Selected Bibliography of Coal Gasification," Bureau of Mines, Report of Investigations 4926 (1952); Bureau of Mines Synthetic Liquid Fuels Program, 1944-1955, Part 1: Oil from Coal, Report of Investigations 5506 (1959).

44. L.L. Hirst, J.A. Markovits, L.C. Skinner, R.W. Dougherty, and E.E. Donath, "Estimated Plant and Operating Costs for Producing Gasoline By Coal Hydrogenation," Bureau of Mines, Report of Investigations 4564 (August 1949).

45. Bureau of Mines, "Justification for Construction of Coal Hydrogenation Plants," Mimeographed, December 22, 1950, File No. 93461, pp. 1-2.

46. Ibid., i-ii.

47. Synthetic Liquid Fuels. Annual Report of the Secretary of the Interior to the Congress of the United States, 1950, p. 5; also Bureau of Mines, Report of Investigations 4770 (February 1951).

48. Synthetic Liquid Fuels. Annual Report.... 1951, p. 9.

49. Synthetic Liquid Fuels. Annual Report.... 1952, pp. 8-9; also Bureau of Mines, Report of Investigations 4942.

50. E.L. Clark, "Bureau of Mines Research on Coal Hydrogenation," Minutes of the Bureau of Mines Coal-to-Oil Advisory Groups Meeting, December 9-11, 1953, Mellon Institute, Pittsburgh, Pa., File No. 57153, p. 127.

51. Synthetic Liquid Fuels. Annual Report. . . .1953, p. 2; Bureau of Mines, Report of Investigations 5043 (April 1954).

52. Bureau of Mines, Report of Investigations 4564, and "Justification for Construction...."

53. "Justification for Construction....," p. 12.

54. Synthetic Liquid Fuels. Annual Report....1951, p. 6.

55. NPC, <u>Subcommittee Report to National Petroleum Council's Committee on Synthetic Liquid Fuels Production Cost</u>, October 15, 1951, p. 19; also <u>Mines Mag.</u> (Colorado), 42, 3 (March 1952), pp. 39-41; "Synthetics Too Costly," <u>Oil and Gas Jour.</u>, 50, 172 (November 8, 1951); R.V. Reeves, "1952: Synthetic Liquid Fuels?" <u>Chem. Eng.</u>, 58, 314 (1951).

56. Ebasco Services, Inc., <u>Coal Hydrogenation Plants: A Review of Certain Elements of the Bureau of Mines Cost Estimates for Synthetic Liquid Fuels, for the Department of the Interior, Bureau of Mines</u>, March 1952, p. 34.

57. W.S.S. Rodgers, Chairman, <u>Interim Reports of the National Petroleum Council, Committee of Synthetic Liquid Fuels Production Costs</u>, January 21, 1952; January 29, 1952; July 29, 1952.

58. See, for example, "How Process Equipment Cost Varied," <u>Chem. Eng.</u>, 59, 191 (February 1952); "Coal Hydrogenation--More Cost Data," <u>Chem. Eng.</u>, 59, 19 (June 1952); pp. 159-61; "Ebasco confirms Bureau of Mines' Estimates of Gasoline from Coal," <u>Chem. Eng. News</u>, 30, 1739 (1952); V.B. Guthrie, "Bureau Whittles Down Cost Estimates for Coal Hydrogenation Plant," <u>Petrol. Proc.</u>, 7, 413 (1952); "Synthetics--Costs Still Too High," <u>Oil and Gas Jour.</u>, 50, 100 (April 21, 1953). "Industry's vs. Bureau's Figures Cost of Hydrogenating Coal to Gasoline," <u>Petrol. Proc.</u>, 7, 321 (March 1952); P.R. Schultz, "Facts, Not Fiction About Synthetic Liquid Fuels," <u>Petrol. Refiner</u>, 31, 88 (March 1952); and W.C. Uhl, "Synthetic Liquid Fuel Industry Still Unlikely for a Good Many Years," <u>Petrol. Proc.</u>, 8, 307 (March 1953). A much larger selection of opinions may be found in W.I. Barnet (comp.), <u>Bibliography of Investment and Operating Costs for Chemical and Petroleum Plants, Bureau of Mines Information Circular 7516</u> (October 1949); E.E. Harton and P.R. Tisot (comp.), <u>Supplemental Bibliography...., Bureau of Mines Informational Circular 7705</u> (January 1955); and E.E. Harton, Jr., (comp), <u>Supplement 2.... (July 1942-June 1954), Bureau of Mines Information Circular 7751</u> (September 1956).

59. See, for example, A.L. Foster, "Can Private Business Compete with Government?" <u>Petrol. Eng.</u>, 24, C3-4 (May 1952).

60. "Justification for Construction....," <u>op. cit.</u>, p. 14.

61. NPC, "Final Report," February 26, 1953, pp. 10-11; "NPC Says Synthetic Liquid Fuels Production is Uneconomic," Chem. and Eng. News, 31, 1090.

62. U.S. Congress, Senate Subcommittee on Appropriations, Hearings on Interior Department Appropriations for 1954, 83rd Cong., 1st sess., pt. 2, May-June 1954, p. 1006; U.S. Congress, House, Congressional Record, 99, pt. 3, pp. 4022-26.

63. U.S. Congress, Senate, Ibid., p. 3355.

64. Synthetic Liquid Fuels. Annual Report....1953, Bureau of Mines Report of Investigations 5043 (April 1954), ii.

65. Interviews with W.C. Schroeder, Mr. Lester Hirst, James McGee, Howard Wainwright, E. E. Donath, Sidney Kattell.

66. U.S. Statutes At Large, 1974, 88, pt. 2, pp. 1878-85.

67. Ibid., p. 1882.

68. U.S. Congress, House, Congressional Record, 95th Cong., 1st sess., April 20, 1977, p. 3330.

Ethan B. Kapstein

4. The Transition to Solar Energy: An Historical Approach

Introduction

The industrialized world is beginning to make the transition from fossil fuels to renewable energy sources. That this transition would someday have to occur has long been known to society's sensitive observers. Those who witnessed the decimation of woodlands, or the exhaustion of coal mines, or the depletion of oil reserves dreamt about the possibility of harnessing sun, wind, and tide.

This article focuses on solar energy research in the United States from 1870 to the present. Interest in this field has been most active when threats of fuel exhaustion were strongest; the 1870s, the early 1900s, and the present day are examples. Yet despite this long period of research, demonstration, and utilization, no history of solar technology has been written. One result is that most of those making current solar policy are ignorant of solar technology's past.

History can be useful to policymakers when it sheds light on the past failures or successes of policies or technologies. Such knowledge may lead us to avoid previous errors, and to capitalize on past achievements. If the following discussion stimulates that sort of action, it will have achieved its purpose.

Solar Pioneers

Solar energy utilization has been an object of research and demonstration in the United States since 1870, when the engineer John Ericsson built his first sun-powered motor in New York City. Ericsson--best known for his ironclad warship Monitor--had a lifelong concern with the depletion of natural resources, and throughout his career he drew plans for harnessing the sun. Ericsson took the opportunity to construct and operate sun motors upon retiring from his engineering practice at age 67.

Ericsson's solar steam engine was composed of three parts—the engine, the steam generator, and the concentration apparatus. The first two were unexceptional, resembling in all essential points a contemporary steam engine. The concentration apparatus, however, which utilized the sun to vaporize water, was unique. It consisted of a water-filled metallic tube which ran the course of a parabolic mirrored trough. The sun's rays struck the parabola and concentrated on the tube. The resulting steam in the tube ran the small engine, but it generated only a fraction of a horsepower [1].

Ericsson had grand hopes for solar energy, despite the fact that his motors were so small. He predicted that an 8000-mile-long, one-mile-wide strip along the equatorial range would ultimately support 22,300,000 solar engines, each having 100 horsepower. As coal neared exhaustion, he foresaw great shifts in international relations: "The time will come when Europe must stop her mills for want of coal. Upper Egypt, then, with her never ceasing sun power, will invite the European manufacturers to...erect his mills...along the sides...of the Nile." [2] The theme of resource depletion and the concomitant shift to equatorial lands would remain the dominant message of solar researchers for decades to come.

Ericsson had a remarkable grasp of the prospects and difficulties that lay ahead for solar energy. He foresaw, for example, that a solution to the energy storage problem was essential if the sun was to be exploited on a large scale. Further, he argued that although solar energy was free, the capital costs of the sun motor would have to drop dramatically in order to compete with coal.

The interest generated by Ericsson is revealed in an article in the <u>1901 Annual Report of the Smithsonian Institution</u>. The author, Cornell engineer Robert H. Thurston, briefly mentioned wind, water, and tidal power before turning to his main subject, solar energy. He wrote that "the solar engine is exciting special interest. It is no novelty, and many inventors have, for years past, worked upon this attractive problem; but probably at no time in the past has the matter assumed importance to so many thoughtful and intelligent men or excited so much general interest." [3]

In the article, Thurston described a solar engine that had been constructed in Pasadena, California. The western states were natural centers for solar research; the climate was right, and few other energy sources were available for jobs such as water pumping. Accordingly, several entrepreneurs ventured west in the hope of building solar motors. The Pasadena engine attracted special attention because of its size. It consisted of an enormous conical mirror, 33 feet 6 inches at the bottom, with a boiler placed at the focal point. The engine produced enough steam to power a 10 horsepower engine, which pumped underground water to nearby farms. The mirror and boiler were placed on an axis, and moved by a clockwork to track the sun. Like the Ericsson motor, the engine was conventional; the concentration apparatus was the exceptional feature.

Thurston had little hope for solar energy until he could devise a successful storage system. The occasional cloud disrupted performance, and the engine could not work, of course, at night. Furthermore, the concentrator was expensive and ungainly. Still, he pleaded for research into alternative energy sources. Failure to pursue this research would have grave consequences, he warned: "The ultimate outcome must be the gradual extinction of our fuel supplies, and if no substitute can be devised... the compulsory retreat of civilized races into the tropics." [4]

Another pioneering group of solar workers in the West were the team of H. Willsie and J. Boyle. Beginning in 1904, they built four solar power plants over five years. The first one was built for an exhibition in St. Louis, the others as ventures in Arizona and California. Willsie wrote in a 1909 Engineering News that "we began these experiments because the engineering properties of sun power development have interested me since 1892, and because Mr. Boyle has properties that would be benefited by a successful sun power plant." [5]

Willsie and Boyle introduced new concepts to solar technology. They did not use expensive reflecting mirrors to concentrate the sun's rays, as did earlier engineers, but instead employed sun-heated water to volatize another liquid, which was then used to power a heat engine.

The solar plant consisted of a solar water heater, a storage tank, a boiler, an engine, and a condenser. The solar heater was a shallow wooden trough tightly covered with a double layer of window glass. The sides and bottom were insulated with hay sandwiched inside the frame. The trough was lined with tar paper, and filled with water to a depth of 3 inches. Sun rays struck the trough and heated the water;

this hot water then went to a storage tank. From the tank, the water circulated about the boiler containing the volatile liquid, either ammonia or sulphur dioxide. The liquid vaporized, powering the engine. Afterwards the liquid would condense and return to the boiler [6].

Willsie and Boyle were, to a certain degree, victims of bad timing. Their systems were getting "bigger and better" just as natural gas was being introduced to the West and Southwest. Cost was also a factor. A Willsie and Boyle solar plant cost about $2500, as opposed to a conventional coal plant that cost approximately $600. This high initial cost, and the availability of cheap natural gas, doomed the solar plant.

Solar research was also being conducted on the East Coast. The Philadelphian Frank Shuman began his work in 1906, after making a small fortune with the invention of a widely used fire retardant building material. His first solar motors were similar in design to those of Willsie and Boyle, but his thoughts turned to the East, not to the West. After several years of backyard experimentation, he felt prepared to seek capital for the construction of his dream: a large solar plant in North Africa. In 1911, he founded the Sun Power Company, Ltd., and went to London in search of capital. With the aid of C.V. Boys and other influential scientists, he raised the necessary funds [7].

Like his predecessors, Shuman was motivated to conduct solar research by a vision of the depletion of coal. He wrote that "sun power generators will, in the near future, displace all other forms of mechanical power over at least 10% of the earth's surface, and in the far distant future, natural fuels having been exhausted, it will remain as the only means of existence of the human race." [8]

Shuman's great sun machine was built at Meadi, seven miles south of Cairo, in 1913. The plant consisted of five sun absorbers which were parabolic-mirrored troughs, 13 feet 4 inches wide and 204 feet long. The troughs were spaced 25 feet apart, so that each would not cast a shadow on the one in front when the sun was low. The sun's rays focused on a boiler--or water pipe--which ran the length of the trough, and the absorbers were set on crescents which tracked the sun [9]. The boilers produced steam as early as 6:30 a.m., and over the ten hour working day, the sun machine created an average of 1,100 pounds of steam at 15 pounds pressure per hour to develop 50 horsepower. The engine pumped Nile water to local fields, and freed thousands of workers from the arduous task of carrying water buckets.

Many distinguished Europeans came to tour the Shuman plant in 1913, and it generated great interest. But with the commencement of World War I, operations ceased, and Shuman returned to Philadelphia, where he died in 1918, before the war was over.

Solar Water Heaters

Solar energy met with its greatest success in the mundane application of heating water for domestic service. A thriving solar water heater industry once existed in the United States, and hundreds of thousands of units operated in California and Florida. The sun provided a simple solution to people who lacked other heat sources.

In the late 1800s, Southern Californians drew warm water by exposing a storage tank to the sun. Wood was in short supply, coal was expensive, and in any case, it took a long time to boil a tubfull of hot water on the stove. A Maryland man who had ventured west, Clarence Kemp, realized that improvements could be made to raise the efficiency of the solar hot water system. He placed the tank inside a white pine box insulated with felt paper, and covered it with a sheet of glass. By employing the "greenhouse effect", high temperatures were quickly achieved. Kemp sold his idea, which he called the "Climax Solar Water Heater," in 1895 to a Pasadena businessman [10].

Enterprising Californians were quick to improve the Climax. In 1898, Frank Walker moved the water tank to the house attic, placing it so the upper portion jutted out from a hole he made in the roof. This upper part was covered with a plate of glass, and at the tank's top he put the hot water outlet. The cold water inlet was placed at the tank's bottom to achieve circulation. Walker sold this system for $50.00 [11].

A major advance in solar water heating technology introduced in 1909 by William Bailey provided the model for solar collectors up to the present day. His major innovation was to replace the tank as the central heating unit with a series of copper tubes. The solar collector was a shallow glass box, lined with felt paper, containing copper tubing coiled across a copper sheet to which it was soldered. The copper sheet and tubing were painted black. Instead of heating a water tank, the sun struck only the pipes, and these carried hot water to a storage tank that was placed above the collector. The bottom of the tank was connected by pipe to the lowest tube in the solar collector, while the top of the tank was connected to the highest tube so that hot water entering the storage would drive the cold water at its bottom through the collector to be heated. In this way, natural circulation-- the thermosiphon principle--was achieved [12].

Bailey called this device the "Day and Night Solar Heater," for there was enough warm water in the storage tank to serve a family all evening. The unit sold well, peaking in 1920, when over 1,000 were installed. Californians bought the systems without government support or tax breaks, but the market for solar heaters eroded as natural gas pipelines invaded the area. In 1926, Bailey sold just 350 units, and by 1930, the solar boom was over. This did not mark the end of the industry, however, since Bailey had sold marketing rights in 1923 to the Solar Water Heater Company of Miami, Florida. An ensuing market in Florida was much larger than that of California, and it grew until the beginning of World War II.

The Florida industry developed slowly during the 1920s, but installations rose each year, and an increasing number of firms competed for orders. The "take-off" occured between 1936-1941, when nearly 100,000 units were installed. Over ten firms competed for the solar heater business. Residential home owners were able to finance their purchase with a low interest Federal Housing Authority Title 1 Home Improvement Loan--the first government incentive for solar. In some government-built housing, the FHA even required that solar water heaters be installed. The majority of units, however, were paid for by the homeowners themselves, with no government assistance [13].

Following World War II, the solar industry--which came to a halt during the conflict due to materials restrictions-- was unable to regain its prosperity. Four factors converged simultaneously to destroy any advantages held by the solar water heater. First, the heater fell victim to falling electric rates as this energy source expanded in the 1950s. Second, since many of the solar units were twenty years old, and were beginning to rust and malfunction by the 1950s, homeowners decided to replace them with new electric water heaters. Moreover, people who were preparing to buy a heater saw the old solar systems, and decided against them. Third, a new force, the large-scale builder-developer, appeared on the housing scene, and the individual homebuyer no longer had the choice of heating equipment to be installed. Developers almost always included electric hot water systems in new homes because of their low capital cost; monthly electricity bills were not their concern. Fourth, and finally, Florida Power and Light mounted a major publicity drive for electrification as capacity expanded. People were encouraged to buy electric appliances, and the advertising worked. By the early 1950s, the solar water heater industry was dead [14].

The history of these water heaters had been largely forgotten. This has partly been the result of propensity of solar technology to "vanish without a trace." Ironically, this factor points to one of the great strengths of solar energy. These technologies can operate quietly and efficiently, without leaving a legacy of pollution, radioactivity, or raped landscapes. The history also points to the fact that simple, inexpensive technologies can be utilized effectively as we seek alternatives to conventional fuels. Through a combination of competition and limited government intervention, a new solar industry could contribute in the near term to the important goal of energy conservation.

University Research

Solar energy received its first major boost in the Northeast. In 1937, the Boston philanthropist, Godfrey Lowell Cabot, endowed Harvard and M.I.T. with over $600,000 each for solar research. Harvard was to study how to increase tree production, for Cabot believed that wood might again become a major energy source; M.I.T. was to study mechanical utilization of the sun. Because this discussion is primarily concerned with solar technology, only the M.I.T. program wil be discussed.

Upon receipt of the Cabot endowment, M.I.T.'s Technology Review wrote that "solar energy is the ultimate source of our fuels--wood, coal, oil, and gas--as well as of power derived from wind or falling water. However, the stores of fuel energy. . .are not inexhaustible. It is therefore of ultimate importance to investigate and develop alternative sources of heating and power." [15] It cannot be overstressed that research in solar energy has coincided throughout our history with those periods when future fuel shortages were most feared.

The first objective of the M.I.T. solar group--led by engineer Hoyt Hottel--was to study in depth the solar heat collector. Hottel's interest, however, was to use the sun for space heating of houses, rather than just for water heating. Solar House I was constructed at M.I.T. in 1939 and in 1941 the now-classic paper on the performance of the solar collector was presented to the American Society of Mechanical Engineers [16].

Solar House I--a small structure with total floor area of 16 by 31 feet--was the first house ever built to rely on an active solar collection system for its space heating supply. The type of collector used was not unlike that used in Florida for hot water heaters. The sun-heated water passed to a storage tank in the basement, giving up its heat to an air duct that passed along the tank. The heated air was then circulated through the building. A large basement storage tank held sufficient hot water to warm the air duct, and thus the house, during several consecutive cloudy days [17].

Prior to 1959, M.I.T. built three more solar homes. Each house became the subject of vast publicity, and visitors came from around the United States and overseas. Solar houses III and IV drew particular attention because families lived in them year-round. So great was the interest in solar-heated homes that in 1950 the New York Herald Tribune declared, "Solar Heating Is Here to Stay!" [18].

Although M.I.T. had demonstrated the efficacy of solar heating, even in New England, the economic picture for the technology continued to be gloomy. The capital cost of the heating system for Solar House IV--a "typical" suburban home--was over $5,500; conventional systems were far cheaper. By 1960, it was apparent that fuel would remain less expensive than solar energy for the foreseeable future, and the Cabot research project gradually dissolved [19].

A second factor responsible for diminishing interest in solar energy was nuclear power. In 1960, the Atomic Energy Commission, the utilities, and the nuclear equipment manufacturers were certain that the United States would soon enter a nuclear era. Scientists, policy-makers, and citizens spoke of a future where electric meters could be thrown away. Sun power appeared superfluous when compared to the might of the atom.

Hottel and his colleagues did not disagree with this assessment of nuclear power. They saw a limited role for the sun, for jobs like house and water heating in sunny, rural communities, where atomic reactors were unnecesary or uneconomic. They thought in terms of a solar/nuclear "partnership." Unfortunately, support for this partnership could not be found [20].

While the M.I.T. program had focused on solar utilization in the United States, during the 1950s another program at the University of Wisconsin studied the sun's potential for the Less Developed Countries (LDCs). This solar research was funded for the most part by the Rockefeller Foundation [21].

Transition to Solar Energy: An Historical Approach   117

At the end of World War II, a number of former European colonies had achieved independence. Many of these were in tropical climates, and most were quite poor. In 1949, President Truman announced a program--named Point Four, since it was the fourth "point" of his inaugural address--which aimed to raise the living standards in these nations. Truman called upon every sector of American society--industry, universities, philanthropies--to get involved with this program [22].

Many scientists responded to the President's call for plans for technical and scientific cooperation. One individual, Farrington Daniels of the University of Wisconsin, stressed the importance of energy to the LDCs, and he believed that the sun might provide a cheap source for states that lacked other fuels. With a grant from the Rockefeller Foundation, he established a Solar Energy Laboratory, and enlisted DuPont engineer Jack Duffie as its director. Initial efforts focused on the development of solar cookers. These simple devices had the potential of saving dung and firewood--the major cooking fuels in the LDCs--for more urgent tasks.

Duffie and his colleagues produced a circular parabolic cooker made from plastic, with a 48 inch diameter and 18 inch focal length. A cross bar held a pot which received the sun's rays with the heat equivilent of a 500-600 watt hot plate. The cooker was covered by aluminized plastic, an inexpensive yet durable reflector [23].

In April 1955, Duffie and consulting engineer George Lof brought 30 solar cookers to Mexico for field tests. Their first impressions were positive. Lof, operating in Monterrey and Terreon, considered his trial results "excellent," and Duffie was also satisfied with the reception outside Mexico City. The people seemed to quickly grasp the advantages of the cooker; they saved time from wood gathering, and money was not spent on kerosene. The cookers were first demonstrated by igniting some paper at the focal point, to reveal the heat. Next, a pot of water was boiled, or some beans cooked. Although the time of preparation was long--generally one and a half hours for a pot of beans--the people seemed impressed.

Within a month of their return to Wisconsin, it became apparent to Duffie and Lof that first impressions would not be sustained. By May, Rockefeller Foundation and Mexican Government scientists concluded that use of the solar cookers was inconsistent, and that many were already broken. The initial field tests ended in failure.

In 1958, the University of Wisconsin's Anthropology Department was invited to study the social problems of introducing the solar cooker, and research was undertaken at the Colorado River Indian Reservation in Arizona. Because the reservation possessed ample sunlight and faced a wood shortage, it was an excellent test site.

The Wisconsin group distributed 41 cookers to Indian families, but soon discovered many problems that hampered its use. The device was easily blown about by strong winds, and the glare off the reflector made use impossible without sunglasses. People were sometimes away in the day, and the cooker could not be used after sundown. Furthermore, the cooker prepared only one item at a time. The cooker was also brittle, and easily broken by children or domestic animals. Finally, uncertainty with the weather did not permit the owners to establish a steady solar cooker habit; rather, the people were expected to use it as frequently as possible, though not necessarily every day [24].

Social impediments to introduction were as great as these technical problems. The Indians were aware that white people didn't use solar cookers, and they wondered why Indians should. A double standard was implied, one the people did not wish to accept. They hoped to progress one day to more modern ovens, but not to a solar cooker.

The cooker experience thus offered the Wisconsin scientists a powerful lesson in technology transfer. Despite their best hopes, solar energy devices could not easily be integrated, even when an apparent "need" existed. The staff of the Solar Energy Laboratory realized that more basic research was needed, and the Rockefeller Foundation supported this activity at Wisconsin until 1965.

Photovoltaics

The 1960s marked a period of decreased activity in research for terrestial applications of solar energy. The Rockefeller Foundation no longer gave grants, and the Cabot Program at M.I.T. was moribund. The U.S. Government, however, was actively supporting the development of light-sensitive cells that converted solar energy to electricity. These were the power cells for our spacecraft, and this technology--photovoltaics--was developing at a number of industrial laboratories [25].

Silicon solar cells were first developed in 1954 at Bell Telephone Laboratories. The Bell engineers were searching for a way to cut energy cost in the South and Southwest, and

Transition to Solar Energy: An Historical Approach 119

they developed the cells to power repeater amplifiers. The cells worked, and solar energy was free, but the cost of making the cells was so high that conventional power sources remained cheaper.

When the Russians launched Sputnik I on October 4, 1957, nobody at first recognized the utility of solar cells for space. American and Soviet planners concentrated on chemical batteries as the primary power source. These, however, only lasted a few weeks, and the utility of the satellite was constrained by the lack of an adequate power source.

By 1958, engineers at the U.S. Army Signal Corps were looking for an alternative, and they incorporated six small groups of silicon cells on Vanguard I--launched March 17, 1958--to power its back-up transmitter. They were unaware of the cell's durability, and the satellite cluttered a radio band for over six years! Two months after the launch of Vanguard I, the Rusians sent up a much larger vehicle which relied on solar cells as its primary power source. From this point on, silicon solar cells were the dominant power source for space vehicles.

Between 1960 and 1973, NASA purchased about 10 million cells at a cost of $50 million. An additional $100 million was spent on the development and construction of the protective glass covers and solar cell array assemblies. Photovoltaic research was supported at the rate of approximately $5 million per year [26].

During the 1960s, a number of firms arose that specialized in photovoltaics. Executives in these companies, however, quickly learned that NASA contracts did not put them on the road to riches. Because of NASA's project structure, production was variable, and firms had to wait for the approval of each contract before producing more cells. This lack of continuous work kept the price of manufacture high, because cost-cutting methods and mass production could not be adopted. Nonetheless, the many advances made in solar cell technology during this period provided the foundation of today's research effort.

The Government Take-Over

This brief review had made it clear that a fair amount of solar energy research was undertaken long before the Department of Energy was founded in 1977. Unfortunately, history has all too often gone unnoticed by energy policymakers. Today's solar programs reveal this knowledge gap.

The federal solar energy program was formed during the panic that followed the 1973 oil embargo. Funding in terrestial applications leaped from the 1960s average of $100,000 per year to $5.1 million in 1973 and $17 million in 1974; for Fiscal Year 1979 the budget is over $500 million [11].

Congress has taken the leading role in support of solar research, but a coalition of environmental and consumer groups, wary of continued emphasis on nuclear power, have pushed their representatives to recognize the solar potential. These efforts were largely responsible for making solar one of the important components of contemporary energy policy. But the exponential growth rate of solar research and development is not without consequences. Administrators have had little time to consider history in formulating contemporary policy. One result, according to an observer of energy policy-making, is the reinvention of several "gold-plated wheels." [27]

The solar space and water heating program offers the clearest example of this tendency. These technologies grab 25% to 30% of the total solar budget for research, development, and demonstration. The result has been the creation of several very expensive collector systems. Yet, if only some time had been dedicated to studying the solar homes designed years ago by George Lof, Harold Hay, Harry Thompson, and others, emphasis might have been placed on less capital-intensive systems.

A glimpse at the history of the Florida water heater industry would have helped us avoid several problems that are appearing today. In areas with hard water, for example, collector pipes are becoming clogged; corrosion and dirt have also hampered performance. These problems were all noted and counteracted many years ago. Furthermore, the price tag on most of the solar water heaters available today make them unattractive for the majority of Americans.

Our disregard for history is not only obvious in domestic policy, but in our dealings with Third World countries, as well. The United States is now promoting the use of solar technology in rural areas of the developing world, and several photovoltaic systems have already been installed. Sadly, this work has been undertaken with little thought about the social stuctures and energy systems that have operated in such areas for untold generations. An awareness of these factors is essential to successful technology transfer. This is the important lesson of the Wisconsin solar cooker experience.

History and Policy

Each day it becomes increasingly apparent that we will have to tap the sun for a portion of our future energy needs. But, in this transitional period, a number of important questions concerning solar policy have arisen. Which solar technologies should be emphasized? What percentage of our energy can be gathered from the sun in the year 2000? In 2020? What portion of the energy RD&D budget should go to solar? These are among the questions being debated today.

History can ease our transition to solar energy in several ways. First, a public awareness of cases where solar utilization has been widespread and successful--such as in Florida--can diminish doubts concerning the technology. Second, by reviewing the experiences of those who used solar technology, problems can be avoided. A number of the common troubles that appear in today's space and water heating units were overcome years ago. Third, a knowledge of history can save the taxpayer money. Many ingenious solar devices were designed in the first half of the 20th century; all this work should be carefully reviewed. Some space and water heating systems, for example, may compare favorably in a cost/benefit analysis to the systems under construction today. Fourth, recent history shows how active solar systems have been stressed in research, rather than passive designs which use wise architectural practice to maximize nature's heating and cooling systems. Passive solar, however, compares favorably with active solar in a variety of settings. Fifth, history can help us direct today's research effort. The problem of energy storage has long been noted as a major barrier to solar utilization; work in this area should occupy a prominent place in the energy budget. History also makes clear the need for more basic research in materials science and other areas. Sixth, and finally, a knowledge of past technology transfer projects in Third World countries can help us to formulate sound aid programs.

History cannot serve as our only guide to the future. We must also rely on new ideas, and be prepared to strike out fresh paths. Yet during the upcoming energy transition, choices will have to be made, and our memories might help us to make the wiser decisions.

References and Notes

1. J. Ericsson, Contributions to the Centennial Exhibition, New York, 1876, p. 562.

2. Ibid, p. 557.

3. R. Thruston, Ann. Rep. Smith Inst., 1901, p. 265.

4. Ibid, p. 269.

5. H. Willsie, Engineering News (May 1909), p. 513.

6. Ibid.

7. F. Shuman, The Generation of Mechanical Energy by the Absorption of the Sun's Rays, Tacokny, 1911, p.3. on early experiments.

8. Ibid, p. 6.

9. A. Ackermann, Ann. Rep., Smith. Inst., 1915, p. 163.

10. K. Butti and J. Perlin, CoEvolution Quarterly, Fall 1977, pp. 1-13.

11. Ibid.

12. Ibid.

13. R. Melicher, et al., Solar Water Heating in Florida, (National Science Foundation, Washington, 1974.)

14. Ibid.

15. Technology Review, June 1938, p. 363.

16. H. Hottel and B. Woertz, Trans. of the A.S.M.E., February 1942, pp. 92-104.

17. Ibid.

18. New York Herald Tribune, August 27, 1950.

19. Hoyt Hottel, interview with the author, September 8, 1977.

20. Ibid.; on M.I.T. see E. Kapstein, A History of Solar Energy Research in the United States, unpub'd MA Thesis, University of Toronto, 1978.

21. E. Kapstein, Chapt. 4.

22. H.S. Truman, Memoirs: Years of Trial and Hope (Doubleday, Garden City, 1956), p. 234.

23. E. Kapstein, History.

24. E. Kapstein, "The Solar Cooker," unpub'd ms, 1979.

25. M. Wolf, Proc. 25th Ann. Power Sources Symposium, 1972, pp. 120-124, gives brief history of photovoltaics.

26. Ibid.

27. W. Clark to M. McCormack, U.S. Congress, House, Comm. on Science and Technology, Subcomm. on Energy Research, Demonstration and Development, National Solar Energy Research, Development, and Demonstration Program-- Definition Report, 94th Cong., 1st Session, July 16, 1975, p. 2.

Part 2

The Future

Michael D. Yokell

5. The Role of the Government in the Development of Solar Energy

Introduction

In this article I discuss what may be an essential ingredient in the transition from non-renewable to renewable energy resources: federal subsidies. In the first section of the article, the economic justification for federal subsidies to solar energy are discussed. In the second section, the current federal solar program is outlined. Next, the appropriate types of federal subsidy programs are presented. Finally, a methodology for striking the proper balance among these subsidy types is offered, and a concluding comparison is made between the current federal solar program and an optimal program.

Should the Federal Government Subsidize Solar Energy?

The federal government should subsidize solar energy for both "first-best" and "second-best" reasons. These terms are defined by Western Neoclassical economists with respect to a normative standard of an idealized, "perfectly competitive" economy [1]. First-best policies are those which could increase social welfare if the economy were perfectly competitive. Second-best policies are those which are designed to increase social welfare in an imperfectly competitive economy.

First-best reasons for subsidizing solar energy include: innovation often involves a public good externality because the social returns from innovation cannot be entirely captured by an innovator; investors in the innovation process are usually risk averse when investing in new technologies (by pooling the risk over a far larger number of individuals, society should willingly support a higher level of investment in risky innovations than private firms or individuals would); private purchasers of new technology are averse to risk of product failure, improper installation, etc. (the private willingness to purchase an innovative product would be raised to optimal levels if risk could be pooled with other members of society); and, capital market imperfections can prevent a purchaser from taking advantage of potential life-cycle savings offered by solar technologies whose first-costs are high.

Second-best reasons for subsidizing solar energy include: large, partially unjustified subsidies to conventional energy technologies currently cause them to be overused relative to solar energy technologies; average pricing arrangements for petroleum [2], natural gas [3], and electricity cause them to be overused relative to marginally priced solar energy technologies; and, large, partially unpriced environmental external diseconomies result from the production and consumption of energy from conventional sources. Conventional energy sources are overused relative to energy from solar technologies, which generally involve significantly less environmental impact. For each second-best problem there is a first-best remedy applicable, as noted below. These remedies do not appear to be politically viable at present. Generally speaking, this is because first-best solutions involve removing existing subsidies that have developed powerful constituencies. Offsetting subsidies may be politically viable because they too have developed or are developing constituencies.

An innovation in production is not merely the invention of a new product or process but requires that the product or process actively begin to penetrate the appropriate market. Innovations per se are thus not patentable. Specific devices or production technologies are patentable, but patent laws do not generally afford great protection to the inventor in a field of technical ferment, both because generic ideas are not patentable and because relatively minor technical changes on a basic technological theme often result in the award of new patents. Thus, a firm which pioneers an innovation may expect to share the profits from the innovation with other firms in the same field. "Fast second" firms entering a new market may capture more of the profits from an innovation than the innovator because the latecomer's R&D expenditures are lower than the innovator's. Moreover, an innovation can result in significant benefits to consumers, which they and not the innovator capture. Thus, on both the production and consumption side, an innovator cannot necessarily capture the full social benefits [4] of his innovation and is, therefore, quite unlikely to invest a socially optimal amount in its development.

Even if an innovator were assured of capturing the full social benefits available from a particular innovation, there is no advance assurance that these will be positive. The entire process of innovation, which includes bringing an invention to market, can be extremely costly and the net social benefits of many innovations are assuredly negative. The private investor, even a large one, is typically risk-averse and requires higher than average expected returns as compensation. Thus innovation does not take place unless high returns are foreseen by the investor. Because the loss from any one successful innovation is a much smaller fraction of society's assets than it is of even the largest firms, society may well be significantly less risk averse to this loss than a private firm would be. Again, less than socially optimal investments are made in innovation. By subsidizing the process, the federal government can raise the level of investment in innovation to socially appropriate levels.

A similar phenomenon can be seen on the consumer side of the market. Any individual innovative product is likely to provide only a relatively small increase in consumer surplus for the typical consumer. Yet if the product fails, the attendant loss of consumer surplus may be large, relative to the consumer's total resources. Even if the statistically expected returns are positive, the typical consumer will be risk averse and consume fewer innovative products than he would if the loss from possible product failure were small. A reduction in perceived risk can be achieved by spreading actual losses over many consumers through a risk pooling mechanism such as insurance. The increase in social welfare attendant upon product failure risk pooling is sufficient justification for this type of arrangement, but it does not necessarily require federal subsidies. Private insurance markets, usually in the form of product service contracts, often provide the essential risk pooling. However, before an actuarial record has been developed (and thus when uncertainty is great) private insurers may be unwilling to enter the insurance market. Therefore, if solar technologies provide positive net social benefits, temporary federal support for solar equipment warranties is desirable.

Solar technologies (with the exception of biomass) do not require much fuel. As a fraction of life-cycle costs, "up-front" costs are higher for solar technologies than for most other energy technologies. For the end-user, financing arrangements are, therefore, more crucial for solar than for other energy technologies. Consider a solar technology whose present discounted costs are lower than those of a competing conventional system (i.e., the solar system is "economic" according to the conventional definition). In today's residential loan market, the solar user may have great difficulty financing the additional costs of a solar system. In retrofit applications, loans are often unavailable, or available at home improvement rather than lower mortgage interest rates. For both the business and residential user, the problem is the lender's use of cash flow rather than economic criteria for evaluating loan proposals. Economically feasible solar systems may have long discounted payback times exceeding the length of typical business loans. If the user has a cash flow problem, which he often does, a solar system becomes an unattractive option.

In both the residential and business cases, federal intervention is called for to improve the functioning of the capital market and to ensure that lending institutions encourage rather than discourage economically feasible solar energy systems.

Before discussing second-best reasons for subsidizing solar energy, it might be useful to ask whether there is anything outstandingly different about innovations in solar energy from innovations in other technologies. The answer is no. Thus for the first two subsidy arguments to be valid, the "traditional" view that private firms invest a less than socially optimal amount in innovative activity must hold (but see Hirschliefer [5], and Kamien and Schwartz [6]). Second-best arguments are more directly concerned with solar energy technologies.

A recent report by Cone et al. [7] catalogues federal subsidies to energy production and estimates their magnitude. The total subsidy was estimated at 217 billion undiscounted 1977 dollars since 1918 [8]. Some of these subsidies can be regarded as first best attempts to improve the functioning of the private energy market, but many of them have little justification other than political expediency. Many examples come to mind [9], particularly from the petroleum industry, which received 60% of the subsidies (not including those to natural gas). Most of these subsidies result in a market price below that which would exist under perfect competition. Flaim [10] has shown this for tax subsidies to the petroleum industry. In this situation, solar energy systems are competing against artificially low-priced conventional energy systems. A first-best solution is obviously to remove the unjustified portion of subsidies from conventional energy sources. This currently appears politically impossible, although progress has been made by largely eliminating the oil depletion allowance. A second-best alternative may be to provide federal subsidies for solar systems to compensate for subsidies to conventional fuels and thus prevent distortion in interfuel competition. While there is no general theoretical proof that an intervention of this type results in a net increase in social welfare [11], it is likely to do so in this case.

In cases where direct solar end-use competes with electricity or natural gas distributed by utilities, a marginally priced good is competing with an average priced good. A similar situation occurs in the market for refined petroleum products under the "entitlements" program. Residential solar heating and hot water systems competing with electric resistance heating systems (now accounting for about 50% of the new market) are a case in point. The end user sees the true marginal cost of the solar system but only an average price for the fuel component of the electric resistance heating system. A first-best solution to this price distortion would be to require utilities to utilize marginal cost pricing and simultaneously tax away any windfall profits that the utility would capture from this pricing mechanism. Since marginal cost pricing has been implemented in only a handful of the nation's utilities, a second-best solution seems to be called for, at least for the time being.

Conventional energy sources are directly responsible for significant external diseconomies. While there has been no definitive work quantifying in dollars the health, property, crop, etc., damages due directly and indirectly [12] to conventional energy related pollution, very rough calculations indicate they are large [13].

Solar energy systems generally have little if any direct pollution associated with them, and their substitution for 10% of conventional energy demand would accordingly provide at least $1.5 billion in direct health-related social benefits [14]. Because pollution is generally unpriced, pollution-intensive products are underpriced and overused. According to the well-known argument [15], a first-best solution to this problem would be the provision of optimal taxes per unit of pollution, rebated to the public on a per capita basis. Under this solution, pollution-intensive energy systems would be automatically penalized compared to relatively pollution-free systems. The average charge required could be on the order of $.20/MMBtu. Since pollution charges have made little political headway in the United States, a second-best solution would be to subsidize low polluting industries

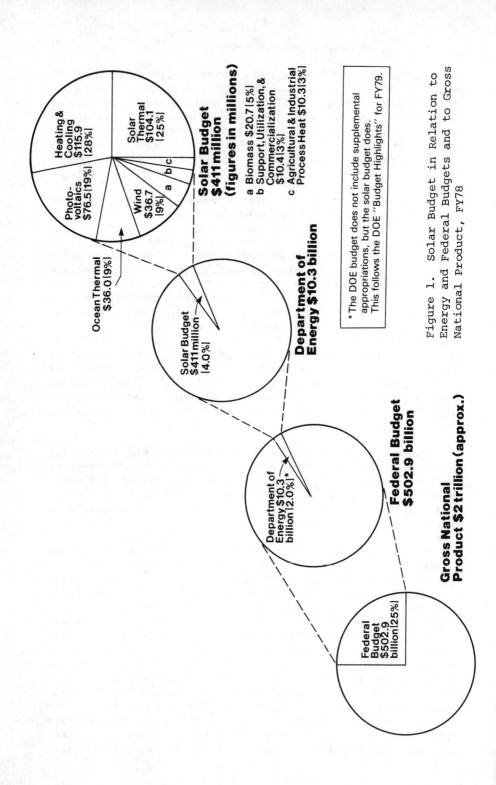

Figure 1. Solar Budget in Relation to Energy and Federal Budgets and to Gross National Product, FY78

which compete in the same market with pollution-intensive industries.

The Current Federal Solar Program

In this section five federal programs for solar energy are discussed: research and development programs; demonstration programs; federal installation programs; purchase programs; and federal tax credits [16].

It is currently the policy of the Department of Energy (DOE) to sponsor significant research and development in the various solar technologies. Figure 1 presents a convenient summary of the budget allocations for solar energy, both relative to other technologies and among solar technologies. The entire Department of Energy solar budget, $411 million in FY78, is only 4% of the total, but this percentage has been growing rapidly.

DOE's solar functions are distributed between the Assistant Secretaries for Conservation and Solar Applications (CS) and Energy Technology (ET) on the basis of technology development status. Solar technologies which are considered mid- to long-term energy supply strategies (solar thermal, photovoltaics, wind energy, and ocean thermal) are located in ET, and presently demonstrable technologies (solar heating and cooling, plus agricultural and industrial process heat) are situated in CS. Thus, most federal solar research and development is located in ET.

Among the principal current and planned ET programs are: technology development in support of the Solar Heating and Cooling Demonstration Program; construction of a 10 MW$_e$ central receiver power plant to test, demonstrate, and produce solar-generated electricity; major procurement of photovoltaic systems and continued research into system cost reduction; continued development in both small wind machines and large-scale, multi-unit wind systems; and conversion of the Hughes Mining Barge into an ocean thermal energy conversion test facility. ET is the principal source of funds for the Solar Energy Research Institute.

The federal government sponsors a program for solar heating and cooling demonstrations. The commercial and industrial portions of this program are administered by the Department of Energy, while the residential portion is jointly administered with the Department of Housing and Urban Development.

In addition to the demonstration program, there is an

active federal program to install solar systems in federally owned buildings. This program is coordinated by the Department of Defense (DOD). DOD also provides a major share of the market for solar electricity.

Numerous other federal entities, from executive departments to national laboratories, have small solar programs. The Small Business Administration has a small low interest (6-5/8%) loan program for solar distributors and installers.

The Energy Tax Act of 1978, passed by the 95th Congress and signed by President Carter, provides for a homeowner tax credit of 30% of the first $2,000 and 20% of the next $8,000 expenditure on a solar or wind energy system for a maximum of $2,000. For solar systems designed to generate electricity, provide hot water, or heat or cool a building, a business is eligible for a 10% tax credit. For solar systems designed to produce industrial process heat or hot water, or for equipment to use, process, or store biomass materials, a business is eligible to receive the existing 10% investment tax credit plus an additional 10% solar tax credit.

Since solar energy systems require large outlays, investment tax credits can be substantial. Accelerated depreciation and rapid amortization are also beneficial to business users of solar energy equipment. They have to be balanced, however, against the absence of any fuel expense which, for conventional energy systems, can be written off directly against income. A cursory examination of the relative impacts of all federal tax laws on conventional and solar systems used in utility applications found no "discrimination" against solar systems, as some solar proponents had claimed [17]. Solar energy is favored, however, as might be indicated by an economically optimal social policy. More than one quarter of the states have passed solar tax incentives legislation. The most generous is California's which allows 55% of the cost of a single-family residential solar system to be deducted from state income tax liabilities, to a limit of $3,000.

Federal Subsidies for Solar Energy

This section presents the policy instruments for subsidizing solar energy. A "policy instrument" can be defined as a generic program. A "program" is any specific instrument whose funding level may be varied independently. This section does not discuss specific solar programs; in the analysis of any specific program its administrative costs and the costs of any side effects must be weighed carefully against the main benefits of the program.

Federal subsidies that would be justified under conventional economic theory for any of the first- or second-best reasons disussed above are not necessarily those that would be suitable for meeting any specified level of market penetration. Much of the literature on federal subsidies for solar energy has attempted to estimate the required type and level of subsidies to meet a specified level of solar market penetration. In the following discussion it is important not to confuse this issue with the "optimal" federal subsidy, which is discussed here.

In Table 1, the economic problems associated with solar energy are listed on the vertical axis and some proposed solutions are listed on the horizontal axis. An "X" indicates that a proposed solution is capable of affecting a problem. The following discussion of proposed solutions refers to the table.

Solution A, direct grants to end-users, compensates for the underpricing of conventional sources of energy and thus is a second-best corrective action. This solution also corrects for the high first-cost barrier and thus compensates for the capital market imperfection, which lowers or eliminates the possibility of end-user borrowing to finance solar systems.

Solution B, income tax credits for end-users, operates in the same way as Solution A except that the subsidy is limited to entities whose taxes are large enough to offset against credits (unless a rebate is provided for).

Solution C, low interest loans to end-users, also compensates for the underpricing of conventional energy. Thus Solutions A, B, and C all affect the second-best problems of solar energy.

Solution D, loan guarantees for end-users, increases the willingness of lenders to loan in an unfamiliar market. Thus capital market imperfections are affected. Lender's risk is reduced below the actuarial level. If this reduction in risk is reflected in lower interest rates, a subsidy is passed on to solar end-users. To a limited extent, therefore, Solution D is capable of affecting second-best problems in the same way that Solutions A, B, and C are.

Solution E, government-provided warranties, or insurance for solar end-users, is designed primarily to reduce the risk to the end-user. The policy that would be optimal from a risk reduction perspective would reduce risk only to its actuarial value, thus charging each end-user an amount to cover

Table 1. First-Best Problems of Solar Energy and Proposed Solutions

Solutions (Proposed) → Problems ↓	A. Direct Grants to End Users	B. Income Tax Credit or Deductions to End Users	C. Low Interest Loans to End Users	D. Load Guarantees for End Users	E. Govt. provides Insurance or warranties for End Users	F. Government Procurement	G. Demonstration Programs	H. Govt. Equity Investment in Manufacturing firms	I. Tax Benefits for Manufacturers	J. Low Interest Loans to Manufacturers	K. Loan Guarantees for Manufacturers	L. Research and Development	M. Federally Funded Training Programs
Private Innovators Cannot Capture Full Social Benefits of Innovation												X	
Individual Innovators are More Risk Averse than Society						X[4]	X	X	X	X	X	X	
Individual End Users are More Risk Averse than Society				X[3]	X		X						X
Capital Market Imperfections	X	X[1]	X[2]	X									

1. The extent to which an income tax credit affects an end-user's cash flow depends on whether a refund is made available immediately when a solar device is installed or later then the tax return is filed.
2. The extent to which low interest loans help overcome capital market imperfections depends on their size relative to the required investment.
3. Government willingness to back a loan may be perceived by end-users as a statement of faith in solar technologies and thus reduce the perception of actuarial risk.
4. Large government procurements can only reduce the risk of investment in innovation if they occur over long periods with guaranteed funding.

Table 1. (continued) Second-Best Problems of Solar Energy and Proposed Solutions

Problems \ Solutions (Proposed)	A. Direct Grants to End Users	B. Income Tax Credit Deductions to End Users	C. Low Interest Loans to End Users	D. Loan Guarantees for End Users	E. Govt. provided warranties or Insurance for End Users	F. Government Procurement	G. Demonstration Programs	H. Govt. Equity Investment in Manufacturing firms	I. Tax Benefits for Manufacturers	J. Low Interest Loans to Manufacturers	K. Loan Guarantees for Manufacturers	L. Research and Development	M. Federally Funded Training Programs
Subsidies to Conventional Energy Sources	X	X	X	X[5]	X[6]	X			X	X	X	X	X
Average Cost Pricing	X	X	X	X[5]	X[6]	X			X	X	X	X	X
Environmental External Diseconomies	X	X	X	X[5]	X[6]	X			X	X	X	X	X

5. Government-backed loans only represent a savings to the end-user if reduced risk of default is passed through to loan purchases in the form of lower interest rates.
6. Government provisions of warranties or insurance only results in an end-user subsidy if such warranties or insurance are provided at less than actuarially required rates.

the cost of random failure of a solar system. Any further reduction of risk provided by Solution E would create a subsidy to end-users and thus function in similar fashion to Solutions A, B, and C.

Solution F, government procurement, is the most direct way of stimulating the market for solar systems. This solution represents a subsidy only to the extent that procurement is for "uneconomic" solar systems. The magnitude of the subsidy is the difference in cost between a cost-effective conventional system and a solar system. In addition to its potential to function as a subsidy to solar energy, government procurement can reduce the risk to innovators if it continues over long periods with funding guaranteed in advance.

Solution G, demonstration programs, is another method for stimulating the solar market. Demonstrations of the technical and economic viability of a technology may reduce the perceived risk to the end-user and thus help to remedy a first-best problem. To the extent that innovations within a technology build upon one another, demonstration of a technology may reduce the risk perceived by innovators of related products or processes.

Solution H, government equity investment in manufacturing firms, reduces the private innovator's risk of failure by limiting his capital investment. In addition, the importance of private failure to capture the benefits of innovation is reduced with increased public participation. Thus, two first-best problems are affected simultaneously. The logical extension of federal participation in innovative projects is complete ownership. Partial or complete federal ownership of business is unpopular, however, and the few examples are special cases [18].

Solution I, tax breaks for solar manufacturers, is another method of subsidizing solar energy and overcoming the hesitancy of manufacturers entering the solar business. Solution I thus ultimately affects second-best problems as do Solutions A, B, and C. The distributional effects may be different, however. Solutions A, B, and C subsidize the end-user's purchase, and the manufacturer benefits by the stimulus provided to the market. In Solution I, however, the manufacturer is subsidized and the competitive mechanism is used to ensure that these benefits are passed through to the end-user. Some of these subsidies would probably remain with the manufacturers.

Solutions J and K, low interest loans and loan guarantees for solar manufacturers or distributors, are similar to tax breaks in that these subsidies reduce the hesitancy of manufacturers in entering the solar business and ultimately lower the price at which solar energy systems can be sold.

Solution L, research and development provided by the federal government, simultaneously affects five of the seven problems outlined earlier. In addition to providing a direct remedy for the inability of private firms to capture the full benefits of innovation, research and development funding reduces the risk of innovations to individual firms and, to the extent that research and development ultimately lowers costs, provides a subsidy to the end-user, though it is not often perceived as such.

Solution M, federally funded training programs for solar architects, engineers, and installers, is an obvious subsidy to solar energy and may have the additional benefit of reducing end-user perception of risk. To avoid overuse by the merely curious, participants should share the costs of these programs.

A number of federal policy instruments and programs are appropriate for subsidizing the development and application of solar energy technologies. How many should be undertaken simultaneously? Each of the first-best problems outlined requires a separate policy instrument if the effect of each solution is to be targetable independently. The second-best problems, however, comprise one fundamental problem: conventional sources of energy are underpriced relative to solar sources. One policy instrument (which could be applied independently to different solar sources to reflect the extent to which their conventional competitors are underpriced) is sufficient to correct all three second-best problems. Thus five policy instruments are required to provide generically optimal subsidies for solar energy. If additional policy objectives are added to a solar program, then additional policy instruments are required to independently target the level at which these objectives are met.

In many cases a number of policies could be used to achieve similar objectives. For example, direct grants to end-users, solar tax credits, and tax benefits for manufacturers all are directed toward the solution of largely the same problems. Selection among competing policies must therefore be based on distributional effects, administrative costs, and public attitudes. Among the three options just mentioned the distribution of benefits varies considerably. It is probable [19] that the benefits from tax breaks for solar manufacturers would be concentrated among fewer recipients than benefits from tax breaks for end-users. It is also probable that providing benefits to solar end-users could be done more cheaply through the existing tax system than by organizing another bureaucracy to administer a direct grant program. The Congress is also usually more inclined to provide special subsidies through the tax system than by direct grants [20]. In view of the foregoing, solar tax credits appear to be economically appropriate (though insufficiently progressive) and politically acceptable method of subsidizing solar energy.

To summarize the results of this section, the broad outlines of an economically optimal and socially acceptable solar subsidy program are presented. First, a major program to compensate for the underpricing of conventional energy sources is required. Subsidies operating on the demand side rather than the supply side of the market are preferable because their benefits are spread more broadly. The major options for providing such subsidies may be ranked in order of increasing pogressivity as follows: tax deductions, tax credits, tax rebates, and below-market-rate loans for solar end-users. If the purported federal objective of progressive income transfers is adhered to, this should be the order of increasing preference. Second, a major program to reduce the end-user's perception of risk is warranted. Here it may be wise to provide different programs for different types of end-users using different technologies. For the homeowner, federal cost-sharing of warranties on solar systems is likely to be the best program. Industrial end-users of large amounts of energy would probably be more influenced by major federal demonstration programs. Commercial users probably stand somewhere in between. Third, a major program is required to reduce the risk of innovation and to compensate for firms' difficulties in capturing the full benefits of innovation. Major federal research and development programs are clearly warranted. In special cases (a solar power satellite system for example), federal equity participation also may be warranted.

The Proper Balance Among Federal Programs to Subsidize Solar Energy

At what levels should components of a broad federal solar program be conducted? In theory, each program should be increased in size until marginal social benefits from the program equal marginal social costs. This condition should be obtained by running federal programs at a level just sufficient to overcome the problems outlined in the previous section. For example, subsidies given to solar technologies to lower their cost relative to subsidized conventional alternatives should be just sufficient to compensate for the subsidies not acting to reduce the price of conventional energy. In practice the problem is considerably more complex. Each solar technology competes in only a segment of the energy market; thus, each solar technology should be subsidized at a different level depending on the average subsidy which affects the price of its competitors. Since some technologies compete in more than one market (SHACOB competes with electric resistance heating in some areas and with natural gas in others, for example), a perfect solution is impossible. Determining the required correction for first-best problems is also difficult. Suppose it is desirable to compensate for private innovators' inability to capture the full benefits of innovation by establishing a federal research and development program. To do this at the "optimal" level would require that we first determine what the investment behavior of innovators would be if they could capture the full benefits of innovation. A research and development subsidy would then be provided which, when added to innovators' private investment in research and development, would equal the level of investment under first-best conditions. A further complication is the possibility that public investment would affect the level of private investment and the two would not be additive.

The foregoing discussion illustrates some of the difficulties with a theoretical economic approach to program funding levels. In the actual policymaking process, the politically desired level of solar market penetration in specific markets is determined and then programs are funded to levels that are judged to stimulate that level of market penetration [21]. Even using this reverse methodology, the problem is complex. Market penetration must be established under varying subsidy types and levels [22]. In cases where the proposed subsidy affects costs directly this is difficult enough; in cases such as research and development subsidies, where the prior impact on costs is unknown, it is an extraordinary problem.

If market penetration is not the only policy objective, a similar but more complex decision scheme must be employed. Such a scheme was used recently by the Department of Energy's "Solar Working Group" in preparing its review <u>Solar Energy Research and Development: Program Balance</u> [23]. In this study, seven choice criteria were used, only one of which was market penetration. Each choice criterion or "value" was assigned a weight for each of three time periods based on the working group's judgment. Different weights were used for each of seven solar technologies, and total "benefit points" were assigned to each technology. Finally, marginal benefits resulting from increases in research and development funding levels were calculated for each solar technology. This phase of the work relied on judging the likely effect of additional research and development funding on the costs of the various solar technologies. The process described led the Solar Working Group to recommend a few major changes in program balance: that solar cooling demonstrations be deferred, that biomass be given additional funding, and that centralized solar thermal electricity be deemphasized. These conclusions depended on the values chosen, their relative weights, the possible effects of various types of research and development funding on future costs, and the effects of cost reductions on market penetration. Judgment was required at each step.

The Solar Working Group/SRI study considered only program balance within the current research and development effort. A more comprehensive study would use a similar methodology to analyze the program balance within other federal solar incentive programs.

Having optimized the balance within programs, the relative merits of each subsidy program need to be analyzed [24]. No comprehensive study has yet done this, but one is surely needed.

Designing an optimal federal solar program requires several phases. First, the program's objectives must be chosen. Second, program elements capable of effecting the objectives must be chosen. This is the simple part of the task and can be done largely by using theoretical tools. Third, the relative funding levels for the elements must be selected. This is a difficult and empirically vexing problem that will remain as long as there is a solar program.

Conclusion

An instructive comparison can be made between the current federal solar energy program, its major economic problems, and the generically optimal program to solve them. Each of the major economic problems of solar energy (investor and end-user perception of risk and underpricing of conventional sources of energy) is now being treated at least partially. Investor perception of risk is treated primarily by research and development funding, though this also compensates somewhat for the underpricing problem. End-user perception of risk is treated by the demonstration program and by the development of federal standards for solar energy systems. Underpricing of conventional energy sources is treated in small part by the solar/wind tax credit.

The major omission in the federal solar energy program is the failure to systematically and thoroughly provide a mechanism to compensate for the substantial underpricing of conventional sources of energy. The solar/ wind tax credit is a positive step, but it treats only two of the eight generic solar technologies in only one end-user sector. A smaller deficiency is the failure to provide a cost-shared warranty program that would reduce the residential user's risk from product failure. When these omissions are corrected, the remaining issues in the federal solar program will be the funding levels and balance among the various programs. The task is to devise and select a mix of programs to maximize social welfare. This task will require continuing research and evaluation of the nation's energy situation, the potential contributions of the solar technologies, and the time-dependent value of those contributions.

My contention is that the value of a given contribution from solar energy technologies increases dramatically through time as conventional fuels become more scarce and expensive. Weingart [25] notes that the world has only three sources of energy large and long lasting enough to provide adequately for mankind's future: fusion, fission with breeders, and solar energy. With the technological, economic, and environmental feasibility of fusion in doubt and the environmental and safety dimensions of breeder reactors a matter of great controversy, solar energy systems may provide the sustainable basis of mankind's energy future. Despite the public clamor for solar energy now, the increasing value of solar energy contributions argues for a long-term emphasis in solar technology research and development funding. For the present, the most significant federal policy would be the provision of subsidies to compensate for the correct underpricing of conventional sources of energy.

References and Notes

1. Ezra Mishan, "A Survey of Welfare Economics, 1939-1959" reprinted from Surveys of Economic Theory, Vol. I, (St. Martin's Press, N.Y., 1967), p. 202.

2. At the time of this writing (April 1979), President Carter had just proposed deregulation of petroleum prices.

3. Under the Natural Gas Policy Act of 1978, price controls on the majority of the U.S. natural gas supply are to be phased out prior to 1987.

4. Edwin Mansfield, et al., <u>Production and Application of New Industrial Technology</u> (Norton, N.Y., 1977). They have attempted to measure the difference between social and private benefits from a variety of innovations.

5. Jack Hirschliefer, "Where Are We in the Theory of Information," <u>American Economic Review</u>, LXIII, 2 (May 1973).

6. Morton Kamien and N. Schwartz, "Market Structure and Innovation: A Survey," <u>Journal of Economic Literature</u> (March 1975).

7. Bruce Cone, et al., <u>An Analysis of Federal Incentives Used to Stimulate Energy Production</u>, PNL-2410/UC-59 (Battelle Pacific Northwest Laboratories, March 1978).

8. This does not include the foreign tax credit, which reduces petroleum price but does not necessarily stimulate domestic production.

9. Tax subsidies for petroleum include immediate expensing for tax purposes of intangible drilling expenses which are economically capital in nature and percentage depletion, which allowed a well to be expensed many times over for tax purposes.

10. Silvio Flaim, <u>Federal Income Taxation of the United States Petroleum Industry and the Depletion of Domestic Reserves</u>, Ph.D. Dissertation (Cornell University, 1977).

11. Richard Lipsey and K. Lancaster, "The General Theory of Second Best," <u>Review of Economic Studies</u>, Vol. XXIV, 1 (Oct. 1956).

12. The term "indirectly" refers to damages caused in the process of manufacturing the inputs to the conventional systems, the inputs to these inputs, etc.

13. Lester Lave and E. Seskin, <u>Air Pollution and Human Health</u> (Johns Hopkins, 1977), estimate that a 58% abatement of particulates and an 88% abatement of sulfur oxides would result in a reduction in health damages of 16.1 billion 1973 dollars or 21.5 billion 1977 dollars (p. 225) from stationary sources alone. Of emissions from stationary combustion systems, electric generation alone represents 63.8% of the particulates and 73.5% of the sulfur oxides; other energy-related sources also exist. Thus, very roughly, 3/4 of the potential

reductions in particulate and sulfur oxide damages (or about $15 billion) are directly energy related (U.S. Department of Energy, Energy/Environment Fact Book, EPA-600/9-77-041, March 1978). Moreover, the Lave and Seskin estimates probably seriously understate pollution damages.

14. Indirect environmental benefits and costs of solar systems also need to be quantified. Research in this area is currently underway. See Michael Yokell, Social Benefits and Costs of Solar Energy: An Economic Approach, INT-78-1 (Solar Energy Research Institute, Golden, Colorado, April 1978); Michael Yokell, The Environmental Benefits and Costs of Solar Energy: An Economic Approach, TR-52-074 (Solar Energy Research Insititute, Golden, Colorado, 1979).

15. William Baumol and W. Oates, The Theory of Environmental Policy (Prentice-Hall, Englewood Cliffs, N.J., 1975), chapter 3).

16. This section relies heavily on Lewis Perelman, ed., Annual Review of Solar Energy, SERI/TR-54-066 (Solar Energy Research Institute, Golden, Colorado, November 1978).

17. Robert Witholder and M. Yokell, "The Effect of Federal Taxes on Solar Systems in Utility Applications," internal memorandum (Solar Energy Research Institute, Golden, Colorado, April 19, 1978).

18. The federal government owns TVA and other hydroelectric projects, uranium enrichment facilities, the Postal Service, and Conrail, for example.

19. In the absence of perfect competition, a portion of the benefits from subsidies to manufacturers would be captured by them. Since there are fewer manufacturers than there are end-users, benefits from subsidies to manufacturers would tend to be more concentrated than subsidies to end-users.

20. Subsidies to agriculture are a major exception.

21. The whole question is quite analogous to the problem of obtaining the optimal level of pollution. From a theoretical point of view, pollution charges should be set at levels that reflect marginal social damages from pollution. Since this level is impossible to calculate, a practical pollution charge would be set to attain a specified (probably non-optimal) level of pollution.

22. The MITRE Corporation and SRI International have both developed market penetration models now being used for planning purposes by the Department of Energy.

23. Charles Hitch, et al., <u>Solar Energy Research and Development: Program Balance</u>, advance copy (February 1978).

24. This two-step procedure assumes that the distribution of benefits among technologies within a subsidy program does not change substantially as funding levels are changed. If this is incorrect, program balance within each program and relative funding levels of various programs must be optimized simultaneously.

25. Jerome Weingart, "Sunlight as a Global Energy Option for Mankind", presented at the Annual Meeting of the International Solar Energy Society (Denver, Colorado, August 1978).

Gregory A. Daneke

6. Forecasting Alternative Energy Futures: The Case of Solar Energy

In the present era of accelerating social and technological change, forecasting, or the assessment of future states of the system, has become integral to policy formulation. Forecasting facilitates the development of preemptive strategies which may divert or rechannel the debilitating effects of change. As noted throughout the new national energy plan (NEP II), forecasting is especially vital to the creation of viable energy policies. William Asher explains that

> Policy makers involved with America's energy issues have learned a great deal about the importance of forecasting in the formulation of energy policy. The electricity shortages beginning in the mid-1960s, the recent oil and gas shortages, and the chronic problem of energy-related pollution all point to both the necessity and the difficulty of energy forecasting.... Energy forecasting is required to prevent actual shortages of energy supplies and, more broadly, to avoid the situations in which policymakers are forced to provide energy very ineffectively. In fact, energy forecasting is a crucial step in the policy for any aspect of energy use that requires substantial "lead times" for development or involves physical limits imposed by resource availability [1].

In recent years recognition has been granted to the potential role of solar energy in meeting present and future energy demands, and thus solar technologies have been the focal point of numerous forecasts and policy studies. By and large, these inquiries into solar energy's potential have proceeded along fairly traditional lines, following procedures and approaches developed for economic and/or technological forecasts. While these studies provide useful information regarding potential "market penetrations" and the comparative viability of certain solar technologies, they have yet to provide an adequate basis for policy formulation. The inadequacies of current forecasts stem from several factors:

(1) the application of outmoded and/or inappropriate forecasting models or approaches;

(2) the imposition of erroneous or partially invalid assumptions, in particular the assumption of a competitive energy market;

(3) the failure to account for the social and environmental factors which might buttress solar commercialization;

(4) the problems generated as a result of the diversity and complexity of solar technologies; and,

(5) the lack of coherent programs from which to draw observations about solar market penetrations.

Of course, these problems are not necessarily equally important in terms of the magnitude of impact upon policy applicability. The first three items are perhaps more central and, while somewhat tautological, item 1 is especially fundamental to policy-relevant forecasting. Matters of approach, however, are the most neglected. Moreover, to the extent that issues of approach arise, the misguided notion seems to prevail that the more sophisticated (the more mathematical) the model, the better the forecast.

This discussion will explore briefly some of the basic issues of forecast approach and how they might relate to the socioeconomic character of solar energy as well as to general questions of policy applicability. I hope to demonstrate that the development of coherent and meaningful solar energy policies requires a teleological approach, one which works backward from the definition of life-quality goals to the determination of policies.

Forecasting: Matters of Methodology

Forecasting is generally thought of as the foretelling of the future. This is a simplistic and somewhat misleading interpretation, unless foretelling implies that the story is often made in the telling. In essence, forecasting is often a matter of creating the future. In this regard, policy forecasting is very different from forecasting the weather for example. That is, predicting rain will not necessarily make it rain, nor will it afford one the opportunity to stop the rain. Policy forecasting, on the other hand, may set forth critical factors necessary to either create a given future or avoid predicted consequences. The policy variability, or course, makes it very difficult to judge the accuracy of policy forecasting because if effective policies are derived and implemented, predicted problems should not occur.

Despite the variability imposed by policy choices, a number of policy applications begin by taking a fairly fatalistic posture. In other words, they attempt to identify relatively stable trends and patterns. Once having forecast a given future, policymakers may take action to change the course of events, but possible actions are tightly circumscribed by future projections. Examples of this tendency are exhibited in municipal service and capital facilities planning in state and local government, market surveys in business and industry, and a multitude of other instances where it is essential to determine a population's needs and wants.

It is important to note that the types of forecasting mentioned above are not merely the result of studying demographic trends or buying behavior, but nearly always involve a variety of critical, yet often hidden, assumptions about human behavior, technological change, and the various interrelations which govern a given phenomenon. For this reason forecasting is often more art than science, although it may appear otherwise. How one pursues this art depends upon the approach chosen. Some approaches can even reduce the fatalistic tendencies mentioned above by expanding the spectrum of social choices. The approach, in turn, depends upon one's philosophical orientation or definition of analytical reality. An understanding of these underlying analytical assumptions is crucial to deciphering the policy implications of a given forecast, yet many analyses draw policy implications without fully comprehending the assumptions maintained by the methodology and their relationship to the phenomena being studied.

A Typology of Philosophical Orientations. To relieve, at least somewhat, the burden of ignorance with regard to the assumptions inherent in forecast methods, policy analysts Ian Mitroff and Murray Turoff have developed a highly useful typology of forecast methodology (Table 1). As they point out, most of the major forecast models are rooted in the philosophies of either Leibniz or Locke.

Table 1. A Typology of Forecast Methodology*

Philosophical Orientations	Concepts of Forecasting	Examples of Methodologies
Traditional Approaches		
Leibnizian	Truth is a function of formal or mathematical relationships	Operations Research, Algorithms [3]
Lockean	Truth is experiential and discernable through statistical relationships	Econometric Models, Regression Analysis [4]
Synthetic Approaches		
Kantian	Truth involves a synthesis of several contributing factors and both formal (theoretical) and empirical components. Analysts seek a consensual perspective.	Synthetic Forecasting, Delphis [5]
Hegelian	Truth is a synthesis of conflicting viewpoints and is thus dialectical	Dialectical Gaming, Cognographs [6]
Non-Traditional		
Singerian	Truth is pragmatic, and it is relative to the goals and objectives that can be engendered for a given system.	Teleological Forecasting, No Standard Methods.

*Derived from I. I. Mitroff and M. Turoff, Spectrum (March, 1973), 62-71.

Leibniz represents the tradition of analytical or mathematical simulation modeling used in engineering and in the burgeoning subfield of operations research (OR). The OR approach works best when the relationships to be modeled are sufficiently clear or simplistic that they can be represented by mathematical equations or algorithms. Thus, OR models usually portray fairly mechanical situations such as electronic interactions, inventory flows, or transportation networks. More ambitious exercises, of course, have emerged, such as the first Club of Rome study by Jay Forrester and Dennis Meadows et al. [2] However, for policy purposes, analytical or mathematical simulation models are most instrumental when one begins with clear goals and objectives or merely has a set of queuing problems to resolve, and these conditions are rarely met.

The tradition of Locke, while a bit more versatile, is only slightly better as a policy-making device. The Lockean tradition adheres to an experiential format in contrast to an analytical one, and thus attempts to base assumptions on the results of empirical observations or data gathering. In short, it is more inductive than deductive. This approach, which has become the basis for most modern economic forecasting, is not without its theoretical content (formal asumptions), however the basic thrust is to identify critical relationships through statistical analysis (largely regression) and to extrapolate these trends into the future.

Prime examples of econometric modeling are found in GNP (Gross National Product) forecasts, population studies, and a number of energy supply and demand investigations. Obviously, these indices are useful in the determination of policy, but once again this type of modeling advances inexorable conclusions based upon a few statistically derived relationships (i.e., they are merely probability statements). Furthermore, econometric models usually rely upon certain hidden assumptions about the viability of various commodity markets as well as on other biases of neoclassical economic theory. Like analytical or simulation models, econometric models work best with well-defined problems upon which there is a high level of consensus about the nature of the phenomena, and with a few concise variables which can be reduced to quantitative terms. Both traditional approaches can be used in conjunction with complex social problems (e.g. pollution, crime, education, etc.) only if one is willing to allow numbers to represent diverse and convoluted social values [7]. But, in fairness to modelbuilders, few claim more than heuristic value for their models, and few (if any) claim an isomorphic correspondence to reality.

As Mitroff and Turoff suggest, recognition of the types of problems discussed above has prompted the exploration of alternative approaches. Some of these approaches follow a <u>Kantian</u> orientation and seek a synthesis of theoretical and empirical elements. Such a synthesis might be achieved by bringing a wide range of policy perspectives together through a Delphi or Fishbowl technique (multiple rounds with consensus oriented feedback) [8]. Other approaches assume a dialectical or <u>Hegelian</u> perspective and attempt to achieve a synthesis through the competition of diverse opinions.

These hybrid, interactive and iterative approaches are also being developed in response to the demand for a more open, less elitist planning process. To a certain extent these new forecast devices are replacing various earlier hybrid systems such as "participatory planning" and "advocacy planning" in the general shift to more "strategic planning" [9].

This recognition of the need for greater public input into planning and management creates an additional avenue for forecast methodology. In those situations in which goals and objectives are unclear, different evaluative interpretations (requiring a multidisciplinary approach) are present (e.g., economic vs. environmental) where reflective reasoning is required, Mitroff and Turoff recommend approaches grounded in the philosophy of Edgar Singer, the American pragmatist. As did John Dewey, Singer held that reality is relative to the goals and objectives of a given system. Forecasts based on this principle would exhibit a teleological or goal-oriented approach. That is, they would begin by positing certain socially viable goals, and then work backward to the structural changes, institutional arrangements, and/or policy commitments required to meet the goals. Though the teleological approach is rarely used in economic or technological forecasting, its potential advantages with regard to policy formulation are significant. William Asher explains:

> One motivation behind the development of normative forecasting is the recognition that forecasting is intertwined with the policy decisions made by forecast-users. If actual trends depend on the policy choices of forecast-users, whose judgment is in theory suspended during their consideration of the forecasts and other data, a forecast that assumes their policy choices is hardly meaningful or welcome. This problem is highly visible in company-level technological forecasting, where the company's ultimate decision

to pursue a particular technology will have an impact on progress in that technology. It is also obvious in providing forecasts to top public policy-makers in areas such as pollution control, where governmental choices on the strictness of antipollution regulation will make a huge difference in the trends that forecasters are expected to project. It is much safer for forecasters to generate a series of conditional forecasts, each based on a policy alternative, than to presume what the policymakers are going to do. A safe and active role for forecasters in this situation is to work backward from desired "end-states" (such as an acceptable level of air pollution) to the policies required to bring them about. [10]

The teleological approach also reduces some of the fatalistic or self-fulfilling aspects of traditional forecasts. Its primary aim is to generate a range of policy alternatives for meeting certain goals. In this way it defines what a commitment to those goals will entail in terms of policy change, and thus it facilitates the collective creation of alternative futures.

Energy Forecasting

Energy forecasting has proven to be a very uncertain enterprise in recent years. Fluctuations in the world price of oil and slowdowns in the ill-fated nuclear industry, coupled with the initiation of conservation have made long-range projections unreliable. More crucial perhaps has been the fact that energy has become a major policy issue. This sudden popularity may serve to expose many of the heretofore hidden assumptions and political influences which have impinged upon energy forecasting.

Initially, one might note that energy projections of recent years have tended to overestimate demand factors and inflate the amount of energy needed in the future. Even moderate 1975 projections predicted a doubling of demand by the year 2000. [11] These predictions have come under challenge by scholars looking more carefully at the various behavioral relationships which make up energy demand. Marc Ross and Robert Williams, for instance, contend that even with a "business as usual" approach to energy conservation, U.S. energy usage will gradually decline to approximately 1/3 the level of growth experienced since 1950 [12]. With rigorous conservation programs, of course, the rate of growth could be cut even more drastically. Indeed, the concept of energy conservation tends to shoot holes in the exponential energy growth models of recent years. In particular, it challenges the once-assumed "iron law" of energy growth tied to economic growth. As Clark Bullard suggests, "studies have shown that it may be possible to gradually reduce the energy required to produce a dollar's worth of GNP to about half." [13]

If a radically different picture of the future were to emerge through the consideration of conservation and alternative energy technologies, it would still be unlikely to have much impact upon decision-makers. The current set of institutional arrangements tends to impose a fairly high discount rate (low future value) on alternative energy systems and on the future generally. As House and Monti explain, futurology in energy modeling is constrained by the following factors:

(1) A politician's survival is tied to the present and not to the future.

(2) Few long-range planning efforts have produced measurable success.

(3) Energy policymakers do not want to tip their hand by involving future contingencies.

(4) Legislatures and bureaucracies are prone to inertia and incompetence.

(5) Effective long-range planning requires considerable expertise and thus considerable capital investment in planning. [14]

Recognizing these constraints does not necessarily dictate the conclusion that forecasting is an insignificant enterprise. The short attention span of policymakers notwithstanding, forecasts are often used to justify foregone political conclusions. Not only are legitimate forecasts used for perverse political purposes, but a number of forecast models are often imbued with a variety of political parameters at the onset of analysis. These predetermined policy guidelines are nearly impossible to detect in the labyrinthine structure of energy forecast models. As one mathematical modeler put it: "We are told to build a model which shows favorably on one particular energy resource, or they will find someone else."

In addition to hiding political influences, the vast complexity of energy models also obscures the remainder of their basic assumptions. Moreover, as model builders spend more time on the superstructure of their models (e.g. the critical operational equations), their assumptions (e.g. the price of a barrel of oil) may become outdated. Asher calls this phenomenon "assumption drag" and suggests that it is a major cause of inaccurate or policy-irrelevant projections [15]. Assumption drag if often compounded by using the results of earlier models as inputs to newer models. This occurs because models naturally become less accurate as the time frame of forecasts is expanded, and using old data adds years to one's forecast [16].

Some of the most perplexing of the hidden assumptions found in many energy models, particularly econometric models, relate to the Lockean tradition mentioned above. Basically these assumptions relate to the existence of an invisible energy market in which all the various resources compete on an equal footing. In essence, by excluding a discussion of governmental incentives, and of qualitative differences between energy sources, a market myth may pervade forecast conclusions. As this discussion will address at greater length below, this market myth is particularly damning to solar energy.

Another factor which is conspicuously absent from energy forecasts is the consideration of the relative social and environmental effects of energy resources [17]. The exception to this is the CONAES study sponsored by the National Academy of Sciences. Including social, environmental, and other somewhat intangible factors would probably combine with the aforementioned problems and topple the methodological utility of the more sophisticated models. In essence, it would raise the level of social and political uncertainty to the point where purely mathematical or econometric approaches would have to be augmented with social interactive techniques.

Solar Forecasts: Procedures in Search of Policies

The problems which plague energy forecasting generally become particularly acute when solar energy is being considered. While the models of operations research and macroeconomics may have some applications to energy generally, they are especially ill-suited to solar forecasting. This lacuna is well exhibited in various major studies of solar energy potential completed in recent years [18].

Solar forecasting is a growth industry. Over one hundred fifty separate studies have been conducted under the rubric of assessing solar potential. Most of these

Table 2. Major Solar Energy* Forecasts (in Quadrillian Btus Displaced).

Study [18]	SOLAR QUADS/TOTAL QUADS		
	Year		
	1985	2000	2020
CEQ	—	24.5/100	45/105
CONAES			
Low Case	0/98	.1/146	—
High Case	3/98	14/146	—
DPR			
Base Case	—	6/95-132	—
Max Case	—	14.3/95-132	—
Mitre/SPURR	.09/85	5/113	—
MOPPS	1.2/95	2.8/117	—
NSF/NASA	.4/117	12/177	109/300
PI			
Base Case	.8/120	11/180	—
Acc. Case	1.4/120	40/180	—
Soft Path	5/95	40/108	70/76
SRI			
Ref. Case	2/99	6/145	11/198
Emph. Case	5/99	15/148	44/204
Low Demand	2/79	7/89	14/102

*Does not include hydropower.

forecasts, however, involve only one particular technology. The list of comprehensive or multi-technology studies is far smaller, though still voluminous. The abundance of sophisticated forecasts notwithstanding, the exact amount of total solar potential remains highly problematic.

In all fairness, while nearly all of the comprehensive studies involve forecasts, they are primarily designed to test for the level of market penetration (commercial acceptance) under a variety of different sensitivities and under a number of different key assumptions. However, their "bottom line" or estimation of total solar output is usually the factor that attracts the most attention. Here the studies differ widely, and thus tend to exacerbate the problems of solar policy development (Table 2).

The basic reason for these diverse results is that the studies assign different values to a number of critical factors or assumptions affecting solar development. The most critical factors might include:

(1) the price of alternative fuels;

(2) total energy demand;

(3) magnitude and success of solar RD&D;

(4) population growth;

(5) conservation;

(6) consumer awareness;

(7) GNP; and

(8) relaxation of institutional constraints.

All these factors, of course, are highly policy-sensitive. In other words, they are relative to one's choice of goals and objectives. Consequently, until these larger questions of policy commitment are resolved it is a little ridiculous to talk about solar potential.

The utility of these studies is thus not one of determining solar's future. At best they provide a reasonable ranking of solar technologies. A workshop on solar forecast models (focusing particularly on Mitre SPURR) sponsored by the Solar Energy Research Institute (SERI) concluded that all the models reviewed had the following limitations:

(1) an inadequate basis in behavioral theory for some of the relationships in the models;

(2) the absence of market data on solar penetration; and

(3) uncertainty about the consistency or accuracy of existing cost and performance data for solar technologies in different stages of the R&D or innovation process. [19]

Thus, the workshop participants concluded that none of the models were adequate to guide policy decisions.

Toward a Teleological Approach

For a forecast of solar energy to be of much use as a policy instrument it cannot very well follow traditional methodological lines. Asking how much solar energy will be available in a certain year is a fruitless inquiry unless one first knows how much commitment a society is likely to devote. If market forces prevail and nonmarket forces (e.g., the environment) are irrelevant, commitment might be determined by private sector interactions. However, in the absence of clear market mechanisms other criteria of social utility become essential and policy should be relative to goals arrived at through social assessment. Whether implicitly or explicitly, most of the existing studies associate market penetration primarily with the price of solar devices. Some even impose a learning curve which makes price the sole determinant of solar competitiveness (e.g. SPURR). If an energy market does not exist, then price is a poor determinant of viability; if viability is relative to normative policy choices, then methodological integrity dictates a teleological approach.

The Non-Market Nature of Solar Energy.
Though solar energy is an ancient energy source, it is largely an outcast in its most recent reincarnation. It has no natural constituency aside from environmentalists and the infant solar industries and their associated services (e.g., plumbers, carpenters, architects). Support from organized labor is certainly logical given the labor intensiveness of dispersed solar technologies, yet this support has yet to fully develop. Meanwhile, solar research, development, and demonstration has only recently received any measurable priority.

Since the turn of the century, government has been a promoter of one energy alternative or another. Oil and natural gas have been major recipients and in recent years nuclear energy has become a focal point of governmental support. One study suggests that approximately $130 billion has been provided in direct and indirect subsidization (Table 3).

In comparison to outlays of public resources for conventional sources, solar energy receives meager favors indeed. As late as the mid 1960s, government funding for solar research was virtually nonexistent. By 1972 funding was only 2 million dollars. In 1977 the figure shot up to $290 million, and future allocations will probably be at least $500 million annually.

To the extent that solar energy receives a share of the federal RD&D budget its future is constrained by the distribution of funding among solar technologies (see Figure 1). While the size of the solar budget has grown, the distribution between large and small solar technologies has remained fairly static. Moreover, little funding has been set aside to discover points of interface between small-scale technologies and community development, existing utility structures, or large-scale technologies. To fund this type of institutional research would require a basic shift in DOE philosophy away from its emphasis on hardware development. The challenging research tasks to be performed involve inquiries into the potential for social change, infrastructure enhancement, and general issues of commercialization. Unfortunately, DOE has demonstrated little interest or aptitude in these areas. The lack of performance standards, licensing requirements, and certification are prime examples of commercial inhibitors. In addition to these factors, the Environmental Law Institute has identified a number unresolved legal issues which constrain solar utilization. These include:

(1) <u>solar access</u> , or "sun rights" (e.g., codes which would prevent the building of units which shade existing collectors;

(2) <u>existing building codes</u> (e.g., limits on types of structures, siting requirements, and hot water levels);

(3) <u>loan and financing problems</u> with regard to solar construction and improvements;

(4) <u>DOE patent policies</u> which inhibit solar innovations under research contracts;

(5) <u>anti-trust</u> issues involving exiting utilities and their involvement in solar energy; and,

Table 3. Estimated Cost of Incentives Used to Stimulate Energy Production*

Energy Source	Time Period	Cost in Billions (1976 dollars)
Nuclear	1950-1976	15.3 - 17.1
Hydro	1933-1976	9.2 - 17.5
Coal	1951-1976	6.8
Oil	1918-1976	77.2
Gas	1918-1976	15.1
TOTAL		123.6 - 133.7

*Taken from: U.S. General Accounting Office, Commercializing Solar Heating: A National Strategy Needed (Washington, D.C.: GAO, 1979) [20].

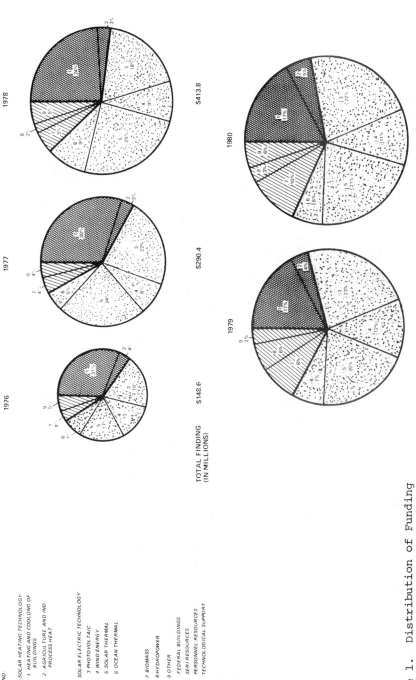

Figure 1. Distribution of Funding Among Various Solar Programs

(6) insurance liability, and warranties involving solar units. [21]

While it may not be advisable for the federal government to establish rigid standards, thereby stifling innovation, it could easily provide model legislation and generalized mandates patterned after the initiatives in states such as California. California already has the highest tax credit available on solar equipment (55%) and provisions for consumer protection (e.g., the Cal-Seal Program); in addition, the SolarCal Council (a state agency) has included the following elements in its "action program":

(1) liberal financing plans;

(2) greater consumer assurances;

(3) broader public information and participation;

(4) job training and development for solar applications;

(5) greater use of solar energy in public facilities; and,

(6) incentives for commercial uses of solar. [22]

Through this program, California hopes to have 1.5 million solar-equipped homes and businesses by the late 1980s.

The removal of institutional barriers is hampered by poor management apparatus in the Department of Energy (DOE), and by the complexity of solar technologies [23]. However, even if management were improved and most of the barriers to solar commercialization were relaxed, it is unlikely that a clear set of market forces would prevail.

A Solar Policy Agenda. Given the non-market nature of solar energy, an assessment of solar potential would begin by exploring those factors associated with its more general social utility. These factors might include:

(1) environmental considerations;

(2) social well-being;

(3) basic political values; and

(4) community self-reliance.

Forecasting Alternative Energy Futures: Solar

While difficult to measure, these factors underscore human relationships to existing and alternative energy systems. The failure to account for these relationships has produced dysfunctional energy policies and programs.

Solar energy's utility hinges on many of these "intangible" factors. For example, solar energy is perhaps the most environmentally benign energy resource [24]. It does not depend upon large amounts of other natural resources (such as air, water, and land) to absorb its residuals. Moreover, solar energy does not contribute to atmospheric pollution which may ultimately cause catastrophic climatic changes.

From a social well-being standpoint, solar systems may free humankind from a portion of the potential health and safety threats which attend continued nuclear development or the economic shocks inherent in easily disrupted centralized utility systems. Furthermore, labor-intensive solar industries could ease chronic unemployment and enhance economic stability.

A variety of basic political values find greater expression in the solar alternative than in the consequences of existing energy systems. These values include: self-determination, deregulation, and personal freedom. Solar installations do not require secret police to guard against sabotage, and they have little value for weapons development.

Finally, the decentralized capabilities of solar technologies seem more compatible with emerging socioeconomic trends of "collective individualism" and "commercial autonomy" than do the capital-intensive alternatives.

Long-range social trends seem to dictate a return to "basic needs" economies, and more atomized community patterns. As mobility is reduced by increased transportation costs, a larger number of smaller and more atomized commercial centers may develop in which the primary energy resource in the sun.

This type of futurology, of course, has had virtually no role in the current practice of solar policy development, and perhaps justifiably so. Nonetheless, the long-term impacts of evolving social values would seem an integral topic when one is talking about solar energy in the year 2000 and beyond. Furthermore, social values are certainly relevant to the discussion of what exactly the nation's solar commitment ought to be.

Initially a thorough policy analysis of solar energy would seem to demand the clarification and incorporation of the types of social values alluded to above. Such social accounting systems are not unique. Paradigms of life-quality accounting are abundant in such diverse areas as water resource management, community capital facilities planning and social service delivery assessments [25]. While not fully operationalized, the Water Resources Council's Principles and Standards for Planning provide a suitable paradigm for multiple account systems [26].

A distillation of social goals and objectives for energy sytems generally could be combined with more basic energy forecasts to form the basis for determining the social requirement for solar energy. This social requirement in turn could be translated into a range of "end goals" or "end states." Even in the absence of an elaborate social assessment, the establishment of hypothetical "end goals" may prove an aid to policy development. As suggested previously, high levels of technological, socioeconomic, and, more importantly, political (or normative) uncertainty require a teleological approach (one which begins with a set of ends and works backwards to engender policy perspectives).

The teleological approach can be pursued through a variety of methodologies; however, scenario development, Delphi, or a combination of both may be useful in speculating about those types of activities required to bring about certain predetermined ends. In essence, surveying a group of energy policy experts should prove instrumental in narrowing the range of policy alternatives. Having established a reasonable range of activities, more standardized techniques of systematic analysis (e.g., optimization, decision analysis, cost-benefit) could be employed to evaluate or rate the various alternatives.

The advantages to the teleological approach are manifold. Initially, a range of end-goals can be considered facilitating a marginal comparison. Much like Zero-Base Budgeting, incremental increases in total solar output could be assessed (1) in terms of social utility minus the level of policy change required, and (2) in terms of the ratio of change to utility.

In addition, market thresholds and other concepts relating to commercialization could be recast in terms of clear objectives and milestones. In this way, perhaps, the digression into mythical market assumptions might be avoided. More important, the bounds heretofore established by arbitrary assumptions are replaced by oportunities for policy change.

Conclusions

In sum, the failure to develop viable policies for solar development is a failure of approach. The history of energy policy is one of producer motivation and thus one of rather conservative approaches to alternative futures. A successful energy transition, on the other hand, will require a focus on the end-users of energy and thus policy-making methods grounded in a broader definition of social utility. Mechanistic modeling, with its emphasis on past experiences, places too many limitations on the future; also, it too often becomes a means of reifying short-term political perspectives. Only through the use of teleological approaches, portraying boldly unique energy futures, will policy forecasting begin to reflect an authentic concern for expanding end-user choices and ultimately improving the prospects of a smooth energy transition. [27]

Notes and References

1. W. Asher, Forecasting: An Appraisal for Policy-Makers and Planners (Johns Hopkins University, Baltimore, 1978), p. 93.

2. D. H. Meadows, et al., The Limits to Growth (Universe, New York, 1972).

3. See: C. W. Churchman, R. L. Ackoff, and E. L. Arnoff, Introduction to Operations Research (Wiley, New York, 1957).

4. Note: S. H. Hyman, "What Is An Econometric Model Anyway?" LSA (Spring, 1978), pp. 8-10 and 16-17.

5. Refer to: H. Linstone and M. Turoff, The Delphi Method and its Application (Elsevier, New York, 1973); also note T. J. Gordon and O. Helmer, Report on a Long-Range Forecasting Study (Rand Corp., Santa Monica, 1964), p. 2982.

6. See: R. O. Mason, "A Dialectical Aproach to Strategic Planning," Management Science (April 1969), pp. B/403-B/414.

7. For a unique set of essays on these and other problems see: M. Greenberger, M. Crenson, and B. Crissy, Models in the Policy Process (Russell Sage Foundation, New York, 1976).

8. For an example of the Citizen Delphi or Fishbowl Methodology, see: T. P. Wagner and L. Ortolano, Testing an Interactive Open Process for Water Resource Planning (Corps of Engineers, Ft. Belvoir, Va., 1976).

9. For a discussion of "strategic planning" see: G. Daneke, Administrative Policy and the Public Interest: An Introduction to Planning and Policy Analysis (Allyn and Bacon, Boston, Forthcoming), Chapter 6.

10. W. Asher, (see 1) p. 213.

11. Note for example: SRI, Fuel and Energy Price Forecasts (Stanford Research Institute, Menlo Park, Cal., 1976); W. G. Dupree and J. S. Corsentino, U. S. Energy Through the Year 2000 (Bureau of Mines, Washington, D. C., 1975); and ERDA, A National Energy Plan for R,D&D (Energy Research and Development Administration, Washington, D. C., 1975).

12. M. H. Ross and R. H. Williams, "Our Present Energy Course," Working Paper (Department of Physics, University of Michigan, 1978); also note: The CONAES Demand Panel, "U. S. Energy Demand: Some Low Energy Futures," in P. H. Abelson and A. L. Hammond, Energy II: Use, Conservation and Supply (American Association for the Advancement of Science, Washington, D. C., 1978), pp. 63-72.

13. C. Bullard, "Energy and Jobs," paper presented at the Conference on Energy Conservation (University of Michigan, November 1977), p. 1; also note D. Hayes, "Post-Petroleum Prosperity" paper presented at the Annual Meeting of the American Association for the Advancement of Science (Washington, D. C., 1978).

14. P. House and D. Monti, "Using Futures Forecasting in the Planning Process: A Case Study of the Energy-Environment Interface," paper presented at the Conference of the American Society for Public Administration (Baltimore, Maryland, April 1979).

15. W. Asher, (see 1), pp. 202-203.

16. Ibid.

17. For a discussion of these considerations see: J. Harte and Alan Jassby, "Energy Technologies and the National Environments," in J. Hollander et al., eds., Annual

Review of Energy, (Annual Reviews, Palo Alto, 1978), pp. 101-146.

18. (CEQ) Counsel on Environmental Quality, Solar Energy: Progress and Promise, CEQ, (Washington, D. C., April 1978). (CONAES) Committee on Nuclear and Alternative Energy Systems, Report of the Solar Resource Group (U. S. Dept of Energy, Washington, D. C, Feb. 1977). (DPR) Domestic Policy Review Integration Group, Status Report: Domestic Policy Review DPR, (Washington, D. C., August 1978). (Mitre/SPURR) The Mitre Corporation, Metreks Division, Solar Energy: A Comparative Analysis to the Year 2020 (The Mitre Corporation, McLean, Virginia, Sept. 1977). (MOPPS) Energy Reearch and Development Administration, Market Orientation Program Planning Study (ERDA, Washington, D. C., March 1978). (NSF/NASA) Solar Energy Panel, National Science Foundation/National Aeronautics and Space Administration, An Assessment of Solar Energy as a National Energy Resource (NSF/NASA, Washington, D. C., December 1972). (PI) Federal Energy Administration, Project Independence: Final Report on Solar Energy Panel (Federal Energy Administration, Washington, D. C., Nov. 1974). (Soft Path) Amory B. Lovins, Soft Energy Paths: Toward a Durable Peace (Ballinger Publishing Co., Cambridge, Mass., 1977). (SRI) Stanford Research Institute, Solar Energy in America's Future: A Preliminary Assessment (SRI, Palo Alto, Ca., 1977).

19. D. Schiffel, et al. "The Market Penetration of Solar Energy: A Model Review Workshop Summary" (Solar Energy Research Institute, Golden, Colorado, 1978), p. 2.

20. These figures do not include the full range of economic incentives. For example, the figures on nuclear do not include the full range of fuel-cycle subsidies (enrichment and waste disposal) or the impacts of accelerated depreciation, normalized accounting and limited liability (via Price-Anderson). For a discussion of these items see: G. A. Daneke, "The Political Economy of Nuclear Development," Policy Studies Journal (Summer 1978).

21. Environmental Law Institute, Legal Barriers to Solar Heating and Cooling of Buildings (ELI, Washington, D. C., 1976).

22. SolarCal Council, Toward a Solar California (State of California, Sacramento, California, January 1979), pp. 23-53.

23. See: Department of Energy, <u>Solar Energy: A Status Report</u> (DOE, Washington, D. C., 1978).

24. J. Harte and A. Jassby (See 17).

25. Note for example: The Technical Committee of the Water Resources Centers of the Thirteen Western States, <u>Water Resources Planning, Social Goals, and Indicators</u> (Techcom, Logan, Utah, 1974); also note the numerous examples in K. Finsterbusch and C. P. Wolf, eds., <u>Methodology of Social Impact Assessment</u> (Dowden, Stroudsberg, Pa., 1977), and Harry Hatry, et al., <u>How effective are Your Community Services</u> (The Urban Institute, Washington, D. C., 1977).

26. See: Water Resources Council, "Establishment of Principles and Standards," <u>The Federal Register</u>, XXXVIII, 1-74 (Sept. 1973).

27. This piece was written while the author was a faculty fellow in the Energy and Minerals Division of the U. S. General Accounting Office. The author thus acknowledges the assistance of Farrel Fenzel and Sharon Farmer. The views expressed, however, are not necessarily those of these individuals or of GAO.

Mark N. Christensen

7. Patterns of Transition: Use and Supply of Energy

Despite an appearance of impasse, the shape of the energy future for the nation is being given clear form by decisions that are being made at the present time. The most visible issues--the focus of headlines and rhetoric on controversies over siting of major new facilities for supplying energy--paint a scene characterized by conflict, delay and impasse. That scene is only half the picture, however, for among the participants within the arena of conflict an increasingly powerful consensus is being established on at least three central points: (1) prices of energy are going up, (2) environmental costs must be accounted for, if not by prices, then by policies, (3) large uncertainties inhere in most of the major factors that are involved in planning for energy. While that consensus serves to delay construction of major new supply facilities, at the same time it provides a powerful incentive to end-users of energy to focus their concerns on improved efficiency in use of energy and in developing "self-help" supplies of energy. That measure of consensus points out with a high probability the general direction that future energy developments will take: (1) increasing constraint on expansion of centralized sources of supply, and (2) increasing emphasis on energy conservation and self-help supplies of energy, both of which will reduce demands for energy in the market.

Because decisions on the parts of end-users of energy have attracted little notice, the direction and the vigor of developments already underway tend to be obscured and underestimated. Decisions of end-users have gone little remarked for several reasons. First, the pattern of decision-making is highly diffuse and decentralized, producing few foci of high visibility. Second, the decisions are largely consensual; so there is little controversy. Third, the decisions lie largely within the scope of effective action and competence of the end-users; political interventions are minimal, chiefly in the form of construction permits, with little political intervention, to this point, in terms of planning or regulation. For such reasons the decisions of end-users create little "news" or political controversy. Consequently, significant changes tend to go unnoticed. Incremental effects are noticed as new data on consumption cause annual (downward) revision of long-range forecasts of energy demands. More attention will focus on these matters as the cumulative effects of the dispersed decision-making become more visible.

The following analysis, then, has two general themes: (1) that at the present time the most important decisions of long range significance are being made not by suppliers of energy and their regulators, but rather, by the end-users of energy, and (2) that the decisions of end-users will reduce aggregate demands for energy substantially below levels that are currently being forecast and will reduce even further demands for marketed forms of energy. To support those assertions the analysis addresses some major elements of consensus within current energy debates and the implications of that consensus for (1) development of large-scale supply facilities, (2) decisions by end-users of energy, and (3) prospects for future demands for energy.

Elements of Consensus

While attention focuses on disputed elements of energy policy, there is nevertheless increasing consensus among informed observers about some basic circumstances which all choices must confront. Each of these factors warrants extensive elaboration, but here we seek only a synoptic overview.

First, there is now a general consensus that prices for energy are going up. Virtually every new source of energy costs more than do existing supplies. The domestically available new sources of energy--frontier oil and gas, synfuels, coal with adequate environmental controls, nuclear power, solar energy--all cost substantially more than do existing supplies. Alaskan gas is expected to cost about 50¢ per therm, double the currently common prices. New electricity generating capacity is expected to cost 50-60 mills per kilowatt hour, compared to 35 mills per kilowatt hour currently in many parts of the country. Potentially "cheap" foreign sources of energy are cheap only so long as they are used in relatively small amounts, on the margin. The nation has recently learned the lesson that among the real costs of our shift to massive dependence on foreign sources of oil are fundamental economic, political, and military vulnerabilities to events in unpredictable, and uncontrollable, parts of the world, not to mention problems of balance of payments, decline of the dollar, stability of the world monetary system, and imperatives for both foreign and military policies and budgets. So, there are no longer any cheap sources of energy in prospect.

The long-term significance--both economic and political--of the prospect of rising energy prices is only beginning to be appreciated. Formerly, when cheap (or seemingly cheap) new sources of energy were available, every expansion of the energy supply system resulted in lower average costs. Now, however, every expansion of the energy system results in higher average costs. Formerly, users of energy had substantial incentive to invest in equipment, facilities, processes, etc., which would consume increasing amounts of energy; under those circumstances the interests of consumers of energy corresponded closely to those of producers who wished to expand supply. Now, however, with prospects of rising marginal costs, users of energy have substantial incentives to make investments that will reduce their future consumption of energy; that interest does not match so well with the interests of producers who wish to expand conventional supplies. That divergence of economic interest also opens up new fields for conflict over policy. The interests of end-users can be expected to affect not only their own specific economic decisions, but also to affect policies at higher levels of aggregation--as, for example, tax incentives. Under these circumstances it seems probable that producers of energy will be caught in a continuing squeeze between declining demands for energy (or at the very least, declining rates of increase of demand) on one hand and cognate changes in public policy instigated by users of energy, on the other.

A second point of consensus is that the user of energy also faces increased recognition that many real costs of energy supply are not counted in market transactions. Real, non-market costs of dependence on foreign fuels were mentioned above. All across the country energy supply facilities have encountered major opposition founded in environmental concerns--air quality, water quality, water quantity, aesthetic impacts, etc. It is clear that environmental concerns are no longer peripheral, but rather that they are central to choices of energy technologies. Here I do not attempt an exhaustive listing of external costs. Suffice it to say that such costs are visible, new costs are steadily being identified, and they will surely be counted in the future calculus of costs of alternative energy options [1,2].

While dispute about how to count environmental costs continues, there is, nevertheless, an effective consensus that those costs must be counted somehow. A measure of that consensus is the National Coal Policy Project [3]. In that project, industrial users of coal took the initiative in establishing a dialogue with environmental groups in order to seek a common ground. Those industrial users are trying not to eliminate environmental controls, but to find a more effective means of planning for environmental controls. It remains to be seen how effective that project can be, but at this point it clearly reflects a major change in attitude on the part of industrial users of coal, a substantial movement toward consensus on the reality of environmental costs.

Environmental problems are so pervasive, however, that even if they can be resolved locally or nationally, irreducible global problems remain. For example, if all the local and regional problems relating to coal are resolved--reclamation of strip-mined lands, transportation, air and water pollution/consumption at the site of combustion, dispersal of acid rains, etc.--there will still remain the global problem of carbon dioxide in the atmosphere and its potential for changing global climate. If problems of radioactive waste disposal should be resolved, there remain the less tractable problems of global proliferation of nuclear weapons or extension of the methods of security and surveillance that are required to maintain security of the nuclear fuel cycle. Such side effects, which are tolerable in small doses, can become intolerable on a large scale. Many of the least tractable enviromental problems of energy supply are functions of the proposed developments.

Finally, environmental problems are increasingly caught up in controversy over how the costs and benefits of economic development are distributed--domestically, globally, and temporally. Many large-scale energy supply projects have the characteristic that the benefits occur in one place or time, while major costs occur elsewhere in space or time--without any adequate mechanisms for compensating the losers. Distribution of costs and benefits poses major political issues, both international and domestic, and is also a major problem for domestic courts, where issues stemming from environmental legislation are a major focus of litigation [4].

In addition to the prospect of rising costs and environmental impacts, a third point of consensus is that there are major uncertainties in most of the factors that affect energy planning by both suppliers and users of energy. There are, or course, uncertainties regarding prices, and environmental impacts and risks (e.g., effects of CO_2, nuclear proliferation), but uncertainty is present in many other dimensions as well: stability of OPEC and its constituent nations; stability of the international monetary system and the institutions of international trade; government policies on energy; future demands for energy; effects of high energy prices and uncertain availability on the overall functioning of the economy; and so on. The seemingly abrupt political changes in Iran in the fall of 1978 illustrate how important changes in the energy supply system can occur rapidly and unexpectedly. The nuclear accident at Three Mile Island, (March-April 1979) and shortages in gasoline supply in the spring of 1979 add to the uncertainties. Prudent observers do not expect these to be the last of the sudden, unexpected, and dramatically important changes. In the face of major uncertainties in many dimensions, increasing numbers of users are developing their plans with an eye to minimizing the shock of unexpected developments.

One important element of uncertainty--the link between energy consumption and economic development--has recently been minimized by developments in both practice and theory. The oil boycott in 1973 demonstrated that abrupt, unexpected curtailment of energy supplies can have dramatic, negative impacts on the economy. That experience raised fears that reductions in energy consumption (= supply) generally would threaten economic development. Both experience and analysis over the last few years, however, have shown that decreases in energy consumption, and increases in efficiency of energy use, if they are foreseen and planned for, can be accomodated without undermining other economic developments. In the last three or four years the industrialized countries have all experienced growth in GNP that outstripped growth in energy consumption, indicating increases in average efficiency of energy use. In fact, the most vigorous economic growth--in Western Europe and Japan--has occured in those countries where the increases in efficiency of use has been the greatest, and where demands for oil have actually decreased. In addition to that practical experience, the National Academy of Sciences recently published an extensive analysis by a group of the nation's most distinguished economic modelers [5], who concluded that if prices rise and investment decisions are made systematically over time, the nation could potentially double its GNP using no more energy than it does now. Both experience and analysis, then, are contributing to an increasing consensus that energy consumption can be restrained substantially through improvements in efficiency, without significant threat to overall economic activity.

The focus here has been on the growing consensus regarding the prospects of rising prices, environmental problems, and uncertainties. This discussion sidesteps the conventional arguments about prospects for depletion of oil and gas, over which there is much arcane dispute. In fact, focus on that item of dissensus serves chiefly to obscure the elements of consensus. Further, even if there were no prospect of depletion of oil and gas, we would be faced with very significant problems of rising costs, environmental impacts, and uncertainties of supply, though the problems would be different in detail. As it is, consensus on the three elements discussed here provides a powerful context for shaping decisions on future energy developments.

Implications for Energy Supply

The consensus described above places substantial obstacles in the path of many proposals for new, large-scale energy facilities. The new facilities are so expensive, and so clearly portend future price increases, that end-users have substantial incentives to look for more cost-effective investments. It is clear that even at current prices it is generally much more expensive to produce a new barrel of oil (or its equivalent) than it is to save a barrel of oil through investments in efficient uses of energy. There are many examples of cost-effective investments in end-use efficiency even at current prices. In California alone, investments in improved efficiency of refrigerators could save the equivalent of 1800 megawatts of electrical capacity at a capital cost that is less than a third of the capital that would be required to produce an additional 1800 MW [6]. For the United States as a whole, investments in improved mileage of automobiles will, by the mid-1980s, save an amount of oil equal to twice the production of North Slope Alaskan oil fields at half the total capital cost [7]. At a time of central concern for inflation, alternatives of that kind are not noticed by informed users of energy, nor will they long be ignored by government policy makers. A systematic search for least-cost alternatives will lead to increasing difficulties for many large-scale supply projects.

As the cost advantages of conventional kinds of energy supply fade, the environmental side effects become more apparent. A systematic comparison of the environmental impacts of various alternative sources of supply (including improvements in efficiency as an effective "source," on the margin) indicates that energy conservation and smaller-scale technologies for use of renewable resources have less environmental impact than most of the conventional large-scale sources of supply [1,2]. A systematic search for least-cost options, which includes environmental and social costs in the accounting, places further obstacles in the path of many conventional proposals.

The distribution of costs and benefits of energy facilities, as well as aggregate costs and benefits, figures strongly in controversies. Political and judicial processes are increasingly concerned with equity and fairness, not just with "utility" and "efficiency." In that kind of social climate, technologies that distribute benefits in one place or time and costs in other places and times will face continuing difficulties. Mechanisms for adequately compensating "losers" of large-scale energy developments have yet to be effected.

Particular kinds of uncertainties occur in most large-scale systems for supply of energy. Foreign sources of oil and gas are subject to social unrest or war in distant lands. New sources of oil and gas from Alaska or the Outer Continental Shelf are developed in particularly difficult or fragile environments and the extended systems of supply are subject to a variety of natural and social surprises, which could interrupt supplies. Labor unrest provides a major uncertainty for domestic coal supplies. Risk of explosion provides special problems for liquefied natural gas (LNG). Nuclear power is plagued with inseparable links to nuclear war or terror. From the end-user's point of view, every one of these supply options is fraught with serious vulnerability to social or natural events that have some reasonable probability of actually occurring. Events have shown that the risks are not hypothetical; many of these kinds of intervention have already occurred—revolutions, terror, sabotage, monetary instability, strikes, wars, boycotts, cartels, and so forth. The formidable administrative and technical capacities of energy industries, or the U.S. government, are no longer adequate to shield users from the impacts. Increasing awareness of uncertainty and vulnerability inevitably leads users of energy to search for alternatives that reduce the vulnerability or, at least, increase chances of surviving unexpected shocks. As costs, impacts, and uncertainties increase, many of the options that are attractive to suppliers of energy become less attractive to the users. The controversies over siting of large energy facilities clearly demonstrate that environmental factors (stretched to include the psychological environment of uncertainty) are now primary considerations, not peripheral matters, in selection among options.

Many large-scale projects are bogged down in a web of liabilities—cost, environment, uncertainty—that does not seem likely to unravel. It is hard to imagine a change of belief or expectation, of cost, or of will that could expedite the conventional kinds of projects on a large scale, overriding all of the problems and objections that attend many of the proposals. On the contrary, it seems likely that each time a major project is approved, the political and organizational capital spent to secure approval is likely to pull the web (or the noose) tighter, making approval of the next project more difficult. The kinds of projects that do secure political approvals will be those that minimize problems of environmental impact, distribution of costs and benefits, and risks and uncertainties. For most of the kinds of large-scale energy projects now in vogue, the prospect is for increasing constraint and difficulty. That conclusion follows, with a high probability, from the measure of consensus that already exists.

Implications for Energy Use

The same consensus that applies a powerful brake to expansion of conventional sources of supply provides a substantial spur to shaping of decisions by end-users of energy. Aggregate statistics on changes in patterns of consumption obscure the effects of the consensus that is shared among informed observers, for the aggregate statistics include the 30% to 40% of Americans who believe that energy problems are some sort of government-industry hoax, and the 40% who don't know that the nation imports oil. On the assumption that the consensus currently shared by the more-informed participants will gradually spread, and on the assumption that perception of the circumstances affects behavior, then the present actions of informed parties should provide a leading indicator of broader public behavior in the future. In fact, the behavior of many informed parties, both individuals and organizations of widely varied character, reveal a consistent, coherent pattern: primary emphasis on conservation of energy, reduction of demands through both increases in efficiency and changes of behavior, and, second, efforts to develop "self-help" sources of energy, in various modes depending on circumstances of the individual or organization.

That pattern of decisionmaking is common to two dramatically different groups: individuals and ad hoc groups that have been called "energy radicals," and the exceptional conservers among business firms, especially energy-intensive industries.

End-users who are concerned to reduce expenditures for energy, and to minimize environmental impacts and vulnerability to uncertainties have a large number of alternative means of achieving those ends, adaptation of both behavior and technology. Specialized journals, such as Energy User News, and, increasingly, the popular press, carry abundant examples of innovative combinations of options by end-users, both individuals and firms, large and small. A listing of kinds of options available provides some flavor of the potential array of adaptations:

Behavioral changes

- Setting back of thermostats.

- Reorganization of work patterns to minimize use of lights, heating, or cooling; e.g., Carter-Hawley-Hale Stores saved about 2.5 million dollars on an annual energy bill largely by reorganizing nighttime cleaning operations.

- Testing for optimum amenity and adjusting heating, cooling, lighting accordingly; e.g., a 65-unit condominium in Florida saved 16% on air-conditioning expense by cycling pumps to be off for ten minutes each hour without the inhabitants noting any decrease in amenity; additional modest design changes saved an aggregate of 50% on electricity consumption.

- Organizing to produce cost-effective energy management; e.g., by 1978 it has become a vogue among large industries to establish corporate-level officers having responsibility to manage efficient uses of energy; Kaiser Aluminum and Chemical, and American Telephone and Telegraph are two prominent examples among many.

Technical Changes

- Insulation; field for cost-effective insulation at present prices is large and new products are rapidly appearing for new kinds of applications.

- Illumination; efficiency is being increased by design changes for daylighting, more efficient devices such as ballasts, adjustments of lighting levels.

- Control systems for sensing, monitoring, and adjusting levels of heating, cooling, and lighting; many new devices and products are appearing on the market.

- Meters not only for electricity, but also for transport of heat by fluids; residential, commercial and industrial experience indicates that feedback of information from meters, especially if linked with incentives, provides a powerful stimulus toward conserving behavior.

- Energy-efficient electric motors; motors having substantially improved efficiencies are beginning to appear on the market.

- Design changes, incorporating various above techniques, to improve efficiency; passive solar houses that require no active heating or cooling are a prime example; 60-70% improvements in efficiencies of refrigerators are feasible; 50% improvements in mileage of automobiles are achievable.

- New industrial processes; for example continuous casting processes can save one step of remelting pig iron; Phillips Petroleum invested two million dollars to modify its process for making butadene, thereby saving 57% of energy consumed in the process, recovering the investment in six months' time of operation.

- Cogeneration in various modes is being pursued by a wide variety of commercial and industrial firms, some using all energy internally, others marketing excess heat or electricity.

- "Self-help" energy, especially solar energy and its derivatives, is being developed not only by individuals and small organizations, but also large-scale industries such as Hewlett-Packard, Monsanto, and Honeywell; in addition, many energy-intensive industries are seeking to develop self-help sources of natural gas or coal (e.g. Allied Chemical, Kaiser Aluminum and Chemical, General Electric).

Implications of organizational changes should not escape note. Coherent social action requires organization. The character of organization reflects purposes and objectives. Large industries are developing corporate-level officers to manage use of energy and development of self-help energy resources. For smaller firms that lack the in-house resources of large enterprises, both newly established firms and existing firms are offering a wide array of management services, systems, and devices for improving energy efficiency and for design and construction of self-help energy resources. Among "energy radicals" informal networks have been established for sharing information and experience, nationally and internationally. For most firms and individuals, the focus of self-help energy is on renewable flows that are generally available; self-help gaseous fuel is a greater interest for larger firms. Further significant organizational developments include formation of a wide array of specialized users' organizations and new professional organizations such as the

Patterns of Transition: Use and Supply of Energy

Association of Energy Engineers. Organizations of this kind are promoting numerous specialized conferences for various kinds of users of energy. Such organizations play an important role in the spread of ideas and of experience.

It remains to be seen just how effective all this activity will be in reducing future demands for energy, but it would be a novel experience indeed if the major accomplishments, in aggregate, occurred before society organized itself for the task of energy conservation.

A relevant observation is that some individuals have achieved reductions in energy consumption of 60-70% using only marginal changes in amenity or behavior. Some firms have effected saving of over 50% in the amount of energy used per unit of output. Many firms in diverse sectors of the economy have realized improvements of efficiency from 30% to 50%. Meanwhile the average decrease in energy consumed per unit of output across various sectors of the economy is generally in the range of 10-20%. Broad-based averages are not likely to be leading indicators of future developments.

Government at various levels is tending to encourage the foregoing pattern of decisions by end-users. A number of cities have adopted or are considering ordinances to encourage energy conservation and development of renewable energy resources. In California the Energy Commission has established new standards for energy efficiency in buildings and appliances, and the Legislature has adopted a tax credit for solar devices. At the federal level, the National Energy Act of 1978 includes tax credits for investments in energy conservation. These governmental actions have attracted some controversy because they affect existing interests, many of which are only indirectly concerned with energy. Nevertheless, controversy on policies for support of energy conservation is much less heated and rhetorical than is controversy over new sources of supply. On the next round of legislation at state and federal levels, the interests of end-users will be represented by increasingly effective organizations.

It should be noted that the improvements in end-use efficiency described above all took place before the major OPEC price increase of January 1979; before the revolution in Iran; before the passage of the National Energy Act with its investment tax credit and proposals for utility rate reform; before the Three Mile Island accident; and before the gasoline supply problems of spring 1979. Utility rate reforms, the elimination of declining block rates, will provide industry with substantial new incentives to reduce consumption of energy and to develop self-help sources of energy.

Whatever the reasons, in the arena of decisions on uses of energy, individuals, business and government are beginning to act in remarkable consort. In a relatively spontaneous manner, without development of eleborate mechanisms, individuals and organizations are developing a coherent pattern of decisions. The reasons, however, should not be hard to identify. The decisions have the effects of: (1) reducing expenditures on energy and minimizing inflationary effects of rising energy costs; (2) minimizing environmental impacts; (3) minimizing social costs, if centralized economic and political planning and decision-making are regarded as costs rather then benefits; (4) effecting a somewhat "fairer" distribution of costs and benefits to the extent that more of the environmental and economic costs and benefits lie closer to the point of use; and (5) reducing vulnerability to a wide array of uncertainties. Coherence in the pattern of action on the parts of such diverse actors stems from consensus about the nature of the problem--or at least about parts of the problem.

It is also worth noting that in the arena of end-uses, social and technical invention are intimately linked together; changes of behavior and organization linked with changes of hardware are vastly more effective than either taken in isolation. Social invention on the part of "energy radicals" is advertised by various slogans such as: self-reliance, voluntary simplicity, integral urban neighborhoods. But social invention is also actively pursued by firms as new kinds of constraints also provide new kinds of opportunities for benefit through cooperation between firms (e.g., cogeneration), between firms and environmental organizations (e.g., National Coal Policy Project), and between firms and the communities in which they operate (e.g., Kaiser Aluminum and Chemical in its relation to the city of Oakland). The character of social invention in the end-use arena stands in marked contrast to the social concomitants of large-scale, centralized energy supply.

Prospects for Future Demands for Energy

If the above analysis provides at least a *prima facie* case regarding the consensus and its implications for supply of and demand for energy, then the implications for future demands are clear: as compared with present conventional wisdom regarding future demands for energy, actual demands will be much lower. Energy conservation through a mixture of technical and social innovation will result in dramatic reductions of demands below currently forecast levels. Development of self-help energy will even further reduce demands for most kinds of marketed energy: low temperature heat, high temperature heat for industrial processes, and electricity. While demands for liquid fuels for transportation can be greatly limited by increases in fuel economy of automobiles, really significant reductions of demand will require social invention--by private choice and public policy together--beyond the kinds discussed here; significant changes in present patterns of location, occupation, and leisure will be needed.

These conclusions hang critically on the assumption that the exceptional energy conservers and activists of today, both "people" and firms, are leading indicators of future energy developments. That assumption is justified in this argument on the grounds that widely divergent actors are beginning to make a common, coherent pattern of decision regarding use and supply of energy; that the coherent pattern exists despite widely divergent values, expectations, hopes, and fears regarding many aspects of the future; that the common pattern follows logically from a consensus that is widely shared among well-informed observers and participants in energy controversies; that because the consensus is widely held by well informed parties having divergent interests, it is likely to be relatively stable through time, to spread and increasingly affect the pattern of decisions of more energy users. That line of argument is, of course, open to critique on many grounds. The question might be asked, "Is this line of reasoning and evidence a less plausible guide to future prospects than are price elasticities (the basis of the more sophisticated economic forecasting techniques) which are based on historical experience under circumstances that are very different from today?"

From the point of view expressed here, the most plausible quantitative forecast of future energy demands known to this author is Scenario II developed by the Demand/Conservation Panel of the National Academy of Sciences' Committee on Nuclear and Alternative Energy Strategies [8]. That projection forecast that aggregate energy demands would rise slowly until about 1990, then begin to decline. The forecast was based on a postulated four-fold increase in energy prices. In the method of projection, price was taken as a surrogate for changes in price together with changes in policy, taste, expectation, circumstance, etc. In my judgment, the changes of behavior postulated in that scenario would be provoked by a two-fold increase in price linked with spreading consensus regarding the environmental problems and critical risks and vulnerabilities inherent in most large-scale, centralized sources of supply. Energy prices have risen substantially since the cited forecast was made (1977), and doubling in the near future now seems probable. If that increase in fact occurs, and if the consensus described here spreads, then aggregate demands for marketed energy could indeed peak about 1990.

Directions for Further Research

Both the evidence and the reasoning presented here flow from a preliminary survey of publicly available information rather than from thoroughly documented, comprehensive research effort. If the argument has some cogency, and because it clearly has implications for policy, it would be desirable to test these conclusions against a more comprehensive investigation of the patterns of decisions of end-users, the factors that affect those decisions, the expectations of further success by the exceptional conservers, identification of leaders and role models in various sectors of society and the economy. Such an investigation could provide critical tests of the breadth and stability of consensus, of the inference that consensus affects behavior significantly, and the inference that the exceptional conservers are in fact leading indicators.

Conclusions

This discussion has focused on the elements of consensus in the energy debate and on patterns of decision by end-users of energy. It argues that the common emphasis on disputed projections of depletion of oil and gas and disputes over siting of energy facilities obscures fundamental choices and developments that are already under way. Consensus on prospects of rising costs, environmental problems, and vulnerability to uncertainties is shaping the energy transition in ways that will constrain expansion of large-scale centralized supply and shift the focus of attention to improvements in efficiency of use of energy and to development of self-help sources. Aggregate energy demands may well peak in the early 1990s and demands for marketed energy may well decline slowly thereafter.

References and Notes

1. J. Holdren, 1977, "Environmental Impacts of Alternaive Energy Technologies for California," in <u>Distributed Energy Systems in California's Future</u>: Interim Report Volume II, HCP/P7405-02, U.S. Dept. of Energy, Washington, D.C., 1977), pp. 1-64.

2. J. Holdren, G. Morris and S. Tannenbaum, <u>Environmental Aspects of Alternative Energy Technologies for California</u>, Report, (1979).

3. Report of the National Coal Policy Project, Summary and Synthesis, Where We Agree (Westview Press, Boulder, Colorado, 1978).

4. B. A. Ackerman, Private Property and the Constitution (Yale University Press, New Haven, 1977).

5. Report of the Modeling Resource Group, Synthesis Panel of the Committee on Nuclear and Alternate Energy Systems, Energy Modeling for an Uncertain Future (National Academy of Sciences, Washington, D. C., 1978).

6. A. H. Rosenfield, D. B. Goldstein, A. J. Lichtenberh and P. C. Craig, Saving Half of California's Energy and Peak Power in Buildings and Appliances Via Long-Range Standards and Other Legislation (1977).

7. D. Yergin, of Harvard Business School Energy Project, quoted in Energy User News (January 29, 1979), p. 6.

8. Demand and Conservation Panel of the Committee on Nuclear and Alternative Energy Systems, "U. S. Energy Demand: Some Low Energy Futures," Science, Vol. 200 (1978), pp. 142-152.

Lewis J. Perelman

8. Speculations on the Transition to Sustainable Energy

The Transition

One hundred years from now, the United States (and most of the rest of the industrialized world) will have completed an epochal historical transition from a society based on depletable, fossil energy to one based predominantly on sustainable, virtually inexhaustible energy. This will be the fourth major energy transition in human history, the previous three being:

- o the transition from hunting and gathering to the harvesting of biomass through agriculture and silviculture;

- o the transition from biomass harvesting to the mining of fossil energy in the form of coal; and

- o the most recent transition from predominant reliance on coal to overwhelming reliance on petroleum and natural gas.

Just as in these past cases, the coming transition to sustainable energy will entail far more sweeping changes than simply the shift from one energy source to another. The transition will entail radical transformations in the panoply of technologies involved in the production, conversion, and consumption of energy; in the structure of institutional, economic, group, and individual relationships; and in the theory, philosophy, values, and goals that define the direction of social behavior [1].

The inevitability of the transition to sustainable energy is cognate to the impending depletion of oil and gas resources. Despite discovery of potentially large petroleum reserves in Mexico, expert opinion differs by at most two or three decades in estimating the time when world petroleum production begins an irrevocable decline. Few if any energy experts would argue with the contention that, one hundred years from now, petroleum as a fuel will be effectively exhausted. Indeed, present thinking is increasingly pessimistic about the future availability of petroleum supplies; a shortfall in world oil supply relative to demand which a few years ago was forecast for the late 1980s [2] has materialized in 1979.

Coal production undoubtedly will be increased to ameliorate the shortfall in petroleum supplies. But coal is almost universally viewed now as a transitional rather than an ultimate replacement for petroleum [3]. Coal's seeming abundance would be rapidly depleted by an attempt to wholly replace petroleum, especially in light of the inefficiencies in converting coal to liquid and gaseous forms. Environmental constraints also are likely to limit the scope of expanded coal production. The high costs, environmental hazards, and physical-chemical limitations of other unconventional fossil fuel resources (oil shale, tar sands, "deep gas," etc.) lead inescapably to the conclusion that no comparably abundant substitute for cheap oil and gas will be found.

The potential for coal and unconventional fossil fuels to replace oil and gas is limited further by the growing demand for these substances as petrochemical feed stocks. The time is approaching rapidly when the demand for hydrocarbon minerals as fertilizer, plastics, chemicals, etc., will make these substances too valuable to burn.

Ultimately, further burning of fossil fuels may be halted by catastrophic climate impacts. Local heating effects from fossil fuel combustion already have become tangible and a substantive environmental issue. If sufficient political anxiety develops over the potential but largely unpredictable modification of climate by buildup of carbon dioxide in the earth's atmosphere, major reduction in fossil fuel use could result; if the feared climate effects materialize, reduction in fossil fuel use will occur, if only as a result of the economic depression that would follow such an environmental disaster [4].

The three major candidates for the principal, sustainable energy source are fusion, advanced nuclear fission, and solar energy (including wind, biomass, and hydropower), probably combined with some use of geothermal energy. Fusion power has yet to be demonstrated even technically possible. Fission power is fraught with technical, economic,

Speculations on the Transition to Sustainable Energy

environmental, and political problems which make its further development increasingly doubtful. Solar energy has a variety of attractive attributes--technical simpliity, diversity of end-use applications, modularity, economic decentralization, environmental benignness, and political popularity--which makes it the most likely basis for the coming sustainable energy system [5].

Although the transition from our present fossil-fuel-based society to one based on sustainable (chiefly solar) energy sources is certain, the specific "path" of the coming transition is uncertain and controversial. The need for major changes in the basic energy systems of the U.S. and other industrialized nations is now widely recognized. But the specific form and direction of the needed energy transformations are the subject of growing public debate and social conflict. Much of the present controversy over energy decisions (both public and private) focuses on the hardware of energy production and conversion, or on the economics of costs, prices, and institutional roles (e.g., the oil industry, electric utilities).

Yet the key to understanding the nature and possibilities of the coming transition to sustainable energy is neither technology nor economics, but is social and individual human behavior. Undoubtedly, important technical developments and even breakthroughs in energy will occur in the next century. But the basic science of energy resources and conversion, and the basic form and function of solar and other sustainable energy technologies are now understood.

My principal concern in this article is not with the energy sources or technology of the coming solar age, but rather with the consequences of the transition to sustainable energy for social structure and behavior. The transition which will occur in the course of the next century or so is likely to be a period laden with intense social conflict, and probably violence. Exactly what kind of society will emerge from this transition period is an extremely speculative question, subject to all sorts of alternative possibilities in light of the unstable nature of the transition process itself. One cannot ignore the potential occurrence of nuclear warfare, catastrophic climate modification, or some other kind of drastic discontinuity which would shatter the underlying structure of human society and lead to new possibilities we hardly can imagine. But within the bounds of an assumption of reasonable continuity under stress, and of a universe of plausible scenarios, I suggest that, at the end of the coming transition, the society residing in the portion of North America we now call the United States is very unlikely to be one which we currently would recognize as a democratic-republican social system.

American society is about to be transformed from one based on fossil energy to one based on solar energy, from an economy of stock to an economy of flow [6], and from a system of limited energy and unlimited power to a system of limited power and unlimited energy. Characterizing the society resulting from this transformation in conventional terms is difficult; certainly it will be different from anything which has existed previously in human history. But granting the inadequacy of conventional language to describe what is both unprecedented and hypothetical, I will suggest that the outstanding social characteristics resulting from the transition to sustainable energy are likely to be feudalism and theocracy.

In describing American society of the late twenty-first century as feudalistic, I must immediately emphasize the limitations on my use of the term. I do not mean to invoke the Hollywood image of monarchies and fiefdoms, knights and castles, etc. I use "feudalism" here to refer to something more generic than the specific social structure of medieval serfdom. Rather, I am concerned with a set of social, political, and economic relationships similar to those which characterized long periods of the history of Europe, India, China, and Japan:

- wealth and power both largely based on land holdings;

- political decentralization ("balkanization" in the modern sense);

- a quasi-steady-state economy; and

- social stratification by caste or class.

This definition of "feudalism" is somewhat unorthodox, though consistent with what some etymologists believe is the origin of the term in the Latin word for "estate." Although the connotations of the term may be misleading, "feudalism" seems the most appropriate conventional word to describe the set of characteristics just listed.

Some readers already may be nonplussed by the suggestion that the above list of characteristics could describe a probable future for American society. A detailed defense of this thesis could be complex and would exceed the scope of this discussion. But a fairly simplistic _a priori_ argument makes the scenario at least plausible.

Feudalism and Theocracy

Throughout history, civilized societies based on solar energy and renewable resources always have been feudalistic societies. This is not merely coincidence. Energy literally is defined as the ability to do work, and therefore is the primal source of all economic activity. In the solar energy society, available energy is proportional to available land. Those who hold the most land hold the most power. Therefore, throughout most of human history, land, power, and wealth usually have been mutually equivalent.

Because of habits of perception and thought, we tend to view the epoch in which we have been living for only about 200 years--the Industrial Age--as the normal human condition, so much so that the sizeable portion of the contemporary human population that lives near or outside the periphery of the industrial realm is considered not only abnormal but in need of salvation. (That what the "undeveloped" may need is salvation *from* industrialization is a proposition only very recently accepted as plausible, and one that still is widely disputed.) Actually in the longer sweep of human history, the Industrial Age is a pretty peculiar state of human affairs, brought about by the benign discovery of energy resources which were based not on the area of territory held, but rather, in a sense, on a hole in the ground.

If you have a fairly small piece of ground which happens to have a big seam of coal or pool of oil underneath it, you can be very wealthy without having much land. You can have considerable wealth and power under these circumstances with only a little land in the right place at the right time. Even very small holes in the ground may yield power (literally, energy available per unit of time) vastly greater and more reliable than that available to the holders of the historically largest estates, kingdoms, and empires. The concentrated power of the propitious holes in the ground is so vast that even the secondary holders of such power--manufacturers, distributors, and others with little or no property but who provide useful services to the "power train" of fossil energy--can and do acquire more wealth and power than most of the old landed aristocracy.

So a social revolution occurs. Wealth and power no longer are based exclusively or even primarily on land but on lucrative holes in the ground and on the attendant technology and organization needed to exploit them. But there is little which is inherently irreversible in the industrial revolution. In fact, when the holes run dry, the revolution can be expected to reverse itself largely though not completely.

Certainly the transition from a society based on nonrenewable resources (area-independent) to one based on renewable resources (area-dependent) must result in major changes in social, political, and economic relationships. And feudalism provides an established model of such relationships adapted to an area-dependent economy.

Theocracy as an attribute of future American society may seem even more surprising than does feudalism, though the two commonly have been linked in history. The full rationale for this speculation also would exceed the present discussion, but the reasons for it have to do with the way societies keep people playing the game of "society" and with the behavioral incentives for people to carry out certain social roles.

The glue that holds our current democratic-republican, capitalist, growth-oriented society together is an amalgam of material economic growth and a distributional lottery. The inducements for individual behaviors necessary to maintaining the identity and integrity of the society as a whole are predominantly and increasingly secular [7]. The liberal separation of church and state reflects a deeper segregation of spiritual from economic affairs which can be traced back to the early roots of the industrial revolution in the philosophy of Francis Bacon [8].

Economic growth cements industrial society by promising ever-increasing rewards to all. In the terms of the game theorist, material growth makes the economy a positive-sum game, in which everyone who participates wins something, albeit that some win more (even much more) than others. As often observed, the monotonically expanding "pie" offers a growing slice to all, regardless of the relative distribution of slices. The result is to give all members of society a stake in maintaining the economic game, and also to ameliorate (though not eliminate) conflicts over the distribution of wealth and income. Economic growth not only has succeeded in defusing the class wars of redistribution which Marx thought would quickly destroy capitalist society, but eventually was embraced by the leaders of the largest Communist countries-- Nikita Kruschev and Deng Xiaoping--as an essential bond to combat the centrifugal forces in those two great confederations [9].

For reasons well understood by behavioral scientists, monotonic economic growth is only a moderately effective conditioner of human behavior, its effects being distributed largely on the basis of what are called fixed ratio or fixed interval schedules of reinforcement [10]. The more powerful component of industrial society's connective glue, and the one exploited most effectively by the United States, is the distributional lottery, based on what the behaviorist calls a variable ratio schedule of reinforcement [11]. The latter schedule is that of the lottery or of any game of chance, providing rewards of varying magnitude at random intervals. A naive observer can see the effectiveness of this arrangement in controlling human behavior demonstrated daily in the city of Las Vegas, Nevada. A more perceptive investigator will detect its workings everywhere else in the U.S. The distributional lottery is incorporated in both our written and our informal constitutions in the form of the Bill of Rights, open elections, anti-discrimination laws, the "common carrier" status in transportation and communication, inheritance taxes, and in the general ethos of "equality of opportunity," at least in the form of equal access.

The essence of the distributional lottery is the promise of rich but uncertain rewards of wealth and/or power for those who are willing to invest in certain economic and political risks. The mythology of equal opportunity is that anyone, with the right combination of luck, perseverance, and talent, can attain the highest levels of success.

The element of luck is crucial to this system and anything which erodes the providential nature of the lottery threatens the very stability of the society. In the mythopoetic vision of our Western movies, cheating at cards seems to be the offense most generously and peremptorily punished by death. In reality, civil disorder in America is most commonly linked to perceptions of bias or discrimination in the treatment of one group or class of people by another, whether black and white, male and female, young and old, or New Left and Old Establishment. Hence the finding of Christopher Jencks's exhaustive study that the causes of inequality in America include a large, irreducible factor of luck understandably produced a sociological sigh of relief [12].

The transition to sustainable energy in the course of the next century promises to dissolve most of the current social glue. The rate of expansion of the economic pie already has been reduced greatly and soon will become effectively zero, at least throughout the long transition period [13]. In the sustainable energy society, one component of the social cement--economic growth--is largely eliminated.

In principle the second component--the lottery--could continue to operate, but the vast reduction in material productivity severely reduces the effectiveness of the lottery as well. The lottery becomes a game not of distribution but of redistribution. Even as the magnitude of the overall stakes is greatly reduced, the risks for the most-well-off are considerably increased. Those who are wealthy and powerful become increasingly inclined to defend their status quo rather than gamble on modest improvements.

But when those at the top of the pyramid refuse to play in a zero-sum lottery, the stakes become too meager to compel the interest of those at the bottom. When the game becomes "penny ante," the serious players drop out. The economic "game" more and more appears to be what has been called "socialism for the rich and capitalism for the poor," especially from the viewpoint of the poor. For example, when the Chrysler Corporation demands one billion dollars in government assistance to protect it from the consequences of negative economic growth, the less-affluent inevitably must feel provoked to make the same kind of demand. But then the "lottery" becomes a "dole"; the productivity-inducing effect of variable-ratio reinforcement is replaced by the depression-inducing effect of noncontingent reinforcement. In simple terms, the rewards for labor, competition, efficiency, and entrepreneurship are diminished or removed altogether.

Before the short-lived age of industrialization and exponential economic growth, societies maintained their integrity over long periods of time, and will be able to do so again after this age's imminent passing. The glue that has held society together during the industrial age is secular, mundane, and material. The traditional social glue which will be restored in the course of the coming transition is sectarian, transcendental, and spiritual. Theocracy provides social order and stability by employing the rewards and sanctions of a dimension of human experience which is free of the physical and social limits of the economic world.

The trade-off between religion and industrialization (based on nonrenewable holes in the ground) long has been observed by historians, sociologists, philosophers, and artists. Marx, of course, viewed religion as antithetical to the inexorable march of materialism. The same view, with a different evaluation, was held by such Romantic poets as Wordsworth, to whom the necessary secularization of industrial society was pathological ("The world is too much with us; late and soon,/Getting and spending, we lay waste our powers..."), and to whom the cure only could be a return to religious spirituality ("...Great God! I'd rather be/A Pagan suckled in a creed outworn...") [14]. Wordsworth's sentiment must have seemed quixotic in a world which then faced another century of exponential growth in "getting and spending," but it now seems an augury of the incipient future.

The coming rise in religious spirituality, resulting from the increased constraint on material growth, leads to a theocratic trend because of its inevitable intrusion into political and economic affairs. Religious influence and secular economic power are complementary; when one grows, the other declines. Symptoms of the coming trend toward theocracy already can be detected:

- explosive growth of charismatic, "born-again" religious sects;

- growth in economic power of both conventional and charismatic religious organizations;

- growth in political impact of theocratic interests (e.g., public aid to parochial schools, and anti-abortion sanctions);

- election of the first non-Italian Pope in four centuries, who already has attacked the political legitimacy of his native government (Poland) and moved to advance Church influence in a constitutionally anti-theocratic nation (Mexico);

- election of the most religiously oriented President in modern U.S. history; and

- insurgent Islamic fundamentalism capped by the theocratic revolution in Iran, directly involving the industrialized nations in the festering religious conflicts of the Middle East.

And signs of the feudalistic tendencies of the coming sustainable energy society also can be seen in current trends:

- the rise of social movements in the U.S. and other industrialized countries identified with concepts such as appropriate technology, environmentalism, zero-growth, decentralization, local self-reliance, libertarianism, and solar energy;

- resurgent geographic, linguistic, and ethnic chauvinism;

- growth of street gangs in urban areas;

- backlash against both illegal and legal immigration, and even intranational migration;

- growing nationalism in provinces of the Soviet Union, Eastern Europe, Yugoslavia, China, Spain, the United Kingdom, Indonesia, and Canada;

- rising impact of terrorism and sabotage;

- decline of voting participation and of public confidence in democratic institutions;

- erosion of monetary institutions with concomitant growth in barter agreements; and

- a growing gap between rich and poor, with reduced inter-class mobility.

Some of the above may be transient events but others mark the beginning of long-term social trends. Nor are the above events completely unprecedented. The trends toward feudalism and theocracy are represented less by novel social phenomena than by the resurgence of historical patterns of social behavior which have been subordinated (but not eliminated) by the economic patterns of the Industrial Age. The coming energy transition constitutes a shift in the balance of influence between the centrifugal and centripetal social forces, and between the spiritual and secular social forces, that always have been in contention. The future social system of North America may not be feudalism and may not be theocracy, but is likely to be more feudalistic and more theocratic than that of the waning era of exponential industrial growth based on fossil energy. These trends do not necessarily condemn future generations to a new form of serfdom. Democratic characteristics may well be preserved in the society of the next century; but only if the feudalistic and theocratic tendencies of the transition to sustainable energy are understood and accounted for.

The Psychology of the Transition

The image of the future I just have presented is not a problem in itself. No particular form of social organization is intrinsically a problem within the framework of its own values. If there is any primal problem for society it is to insure that human beings continue to exist, and undoubtedly there are a variety of ways of organizing future societies with different types of resources and production systems which would permit human beings to endure; of these, feudalism/theocracy has strong historical precedent.

Nevertheless, from the perspective of current social values, some will find the above image of future society

Speculations on the Transition to Sustainable Energy 195

threatening or repugnant. Were this a normative discussion, I might express some personal aversion for the future society I have described. But in this discussion I am viewing the situation essentially as a disinterested observer, and I am saying that the feudalistic, theocratic scenario is the probable outcome of existing trends.

What may be intrinsically problematic about my scenario is not so much the end-state as the transition process itself. The degree of social stress and conflict during the coming transition period has sufficiently great destructive potential to constitute a serious problem. And the key to understanding this problem and hopefully finding appropriate solutions to it is to focus attention on the nature of human behavior. This leads to an observation which I gratefully attribute to the great anthropologist Gregory Bateson. In addition to his anthropological studies, Bateson--one of the few truly transdisciplinary scholars of our age--did seminal work in the field of psychotherapy, particularly in two principal areas: the problem of schizophrenia, and the problem of alcoholism and other forms of addiction.

In a lecture I attended in Colorado a few years ago, Bateson made an observataion which struck me as a very powerful one, and one which has dwelled in my mind ever since. Bateson was ruminating about the behavioral roots of the energy crisis when he confessed that the thought recently had occurred to him that the cure for addiction was schizophrenia. In a discussion of psychotherapy this observation would have been unusual; in a discussion of society's energy problem it was extraordinary.

Bateson compared our energy predicament with addictive behavior such as alcoholism. The analogy of petroleum use with drug addiction is a popular one, depicted graphically by Conrad in his famous cartoon showing Uncle Sam "shooting up" with a hypodermic needle shaped like a barrel of oil. The analogy is actually a completely appropriate one; our collective use of oil and gas has all the classic characteristics of alcoholism or narcotics addiction, including both physical dependency and psychological dependency.

However, Bateson went on to note that such addictive behavior is very hard to change while it is running its course until it reaches a critical point which alcoholics call "hitting bottom." This is the critical moment when the destructive consequences of the addictive behavior are sufficiently conscious and tangible that it becomes possible for the addicted individual to reorganize and to pursue a new direction (though this may not actually occur). And, Bateson said, the collective addiction of our society probably requires a similar kind of critical break before a cure can begin. Furthermore, Bateson suggested that the process of hitting bottom and curing addiction was very much like the process of schizophrenia.

According to Bateson and his followers, the essential etiology of schizophrenia is based on what he called "the double bind" [15]. The double bind is intense existential paradox. Conflict-laden situations become paradoxical as inconsistent outcomes or contradictory messages simultaneously demand some action and prohibit it. The resulting stress cannot be relieved by rational action and leads to a state of profound disorientation clinically diagnosed as schizophrenia. A clinical example illustrates the process:

> A young man who had fairly well recovered from an acute schizophrenic episode was visited in the hospital by his mother. He was glad to see her and impulsively put his arm around her shoulders, whereupon she stiffened. He withdrew his arm and she asked, "Don't you love me anymore?" He then blushed, and she said, "Dear, you must not be so easily embarrassed and afraid of your feelings." The patient was able to stay with her only a few minutes more and following her departure he assaulted an aide and was put in the tubs. [15]

This interaction poses a host of complications and paradoxes for the schizophrenic victim, but essentially they come down to this: "The impossible dilemma thus becomes: 'If I am to keep my tie to my mother, I must not show her that I love her, but if I do not show her that I love her, I will lose her.'" [15].

The schizophrenic state is commonly considered crazy or insane, and certainly can become dangerous and/or self-destructive. However, under some circumstances the same state of consciousness may be illuminating, and perhaps therapeutic. There are numerous stories of people whose schizophrenic-type experiences have led to a creative reorganization of living and a more productive level of existence. Certain religious and spiritual experiences may be of this type. Schizophrenia may be intrinsically neither a bad nor a good thing, but is probably a powerful process by which human behavior attempts to adapt to highly conflict-laden circumstances. Bateson has suggested that the addicted person and perhaps society have to experience a schizophrenic break in order to reorganize and pursue a new adaptive direction. If the hypothesis is valid, the coming transition to sustainable energy may be marked by both the promise and the peril of social schizophrenia. In the balance of this essay, I want to explore the latter thought further.

Conflict

The observation that the coming energy transition will be marked by intense social conflict may not seem particularly surprising. Conflict over energy already can be seen daily in the Congress, in international relations, and in the relations among oil companies, public utilities, labor unions, environmental and consumer organizations, and other institutions and groups in our society. However, the expression of conflict so far has been largely rhetorical and political, and the degree of conflict has been fairly demure and peaceful. Although a few people have died from fights in gasoline lines or from attacks on truck drivers, social conflict of the emotional and physical intensity of the 1960s has not yet been seen in relation to the energy crisis. This benign situation is unlikely to last much longer. The fuel crises of 1974 and 1979 required the mobilization of National Guard units to deal with violence in the trucking industry; the latest major electricity blackout in New York resulted in a plague of looting and other crime. These energy crises lasted only a few days or weeks. The perennial energy supply problems of the 1980s and 1990s are unlikely to entail any lesser degree of social conflict and disorder.

Conflict occurs in different degrees and kinds. Some kinds of conflict result from emotional rancor or simple misunderstanding and can be resolved by communication, reasoning, or simply the passage of time. More intense conflict in the distribution of resources or the structure of relationships may require intense negotiation and ultimately some sort of compromise to be resolved.

But some kinds of conflict are intrinsic, irreconcilable, and absolute. Some conflicts cannot be resolved through compromise. If I am committed to building a dam across a river and you are committed to maintaining the river in its pristine condition, building the dam halfway does not solve the problem. Some kinds of conflict require a winner and a loser. There are only two ways to resolve such absolute conflicts: either combat of some form, or a profound redefinition of reality which renders the conflict irrelevant or imperceptible. Such absolute conflict has been labeled "paradigm conflict" by the science historian Thomas Kuhn [16]; though the concept of paradigm conflict derived originally from the sociology of science, the same concept applies to social epistemology in general [17].

The conflict which will mark the coming transition to sustainable energy will be at this profound level of paradigm conflict. The form of the beginning polarization of social paradigms already can be seen. Amory Lovins coined the terms "hard and soft energy paths" to describe the overall paradigm conflict [18], but this general dialectic has three important components:

- technocracy v. humanism as competing paradigms for the organization of production;

- Freedom A v. Freedom B (to be defined) as competing paradigms for the organization of consumption; and

- centralization v. decentralization as competing paradigms for the organization of social structure.

As Daniel Bell noted a few years back [19], we live in a society whose basic organizational, political, and economic structure increasingly tends toward technocracy, that is, the rule of institutions by a technical elite--engineers, economists, statisticians, scientists, management analysts, and so forth--who are narrowly trained and specialized in specific functions; who seek to base institutional decisions on logic and quantification; and who view social institutions as social machinery. Technocrats increasingly supplant the decision-making prerogatives of such democratic insitutions as legislatures, elected executives, stockholders, union members, and voters. "The post-industrial society involves the extension of a particular kind of rationality associated with science, technology, and economics. When applied to politics, this rationality becomes technocratic, and inevitably creates a populist reaction." [20].

The humanistic backlash to the reductionist rationality of technocracy has begun. The domestic conflict over the Vietnam War was a precursor of a continuing and ingravescent struggle between technocracy and humanism in the remainder of this century. The theocratic revolution in Iran was another augury of this growing conflict. The disaster at Three Mile Island has become the most important (though not the first or only) stimulus to a growing domestic battle over the basic paradigm of technological and economic decision-making.

The second dimension of conflict concerns the concept of "freedom." In the past few years, I have become aware of an interesting pattern in certain political controversies

Speculations on the Transition to Sustainable Energy

generally related to large-scale capital investments, whether public or private. There is little secret that such large-scale projects increasingly have been constrained by political and regulatory opposition associated with the expected environmental, social, or economic impacts of the proposed projects. Regardless of whether the projects involve pipelines, power plants, highways, dams, airports, refineries, mines, or nuclear dumps, a recurrent theme seems always present.

Once these large-scale projects become embroiled in controversy, their proponents at some point will accuse the opposition (commonly labeled as "environmentalists" or more recently "no-growth advocates") of attempting to deprive the American people of their "freedom." This thought typically is expressed something like this:

> The (environmentalist/no-growth) minority who are obstructing this (oil pipeline/power plant/freeway/ dam) do not represent the great majority of the public whose economic wellbeing depends on its immediate completion. Every day this project is delayed is costing the (taxpayer/ratepayer/stockholder/consumer) X million dollars. If this project is not built, people will be forced by this radical minority to change their lifestyles in a way that would be contrary to everything America has stood for. We do not believe the majority of the American people want to give up the personal freedom and prosperity that (the automobile/electric power/the single-family house) has made possible. The (environmentalist/no-growth) minority are elitists who already have these things and now want to close the door behind them, depriving the poor of the freedom to enjoy the benefits of affluence....

And so forth. I have not quoted a specific source here because such statements appear in public hearings, news conferences, and advertisements almost daily. The opponents of large-scale capital projects have evolved an equally typical complementary argument:

> The Company (oil company, electric utility, Corps of Engineers, etc.) is concerned with its own (profits/ power) and not with the public interest. The Company serves the interests of (Big Banks/Big Business/Big

> Bureaucracy) rather than the interests of the
> People of this (community/state/nation/planet). This project is designed to benefit the
> rich and powerful at the expense of the (poor/
> disadvantaged/workers/future generations).
> The project will deprive them of the freedom
> of (a healthy environment/low prices/a safe
> job/community control/ethnic identity/wilderness). This project would destroy (a unique
> culture/environmental stability/scenic beauty/
> an endangered species). Monetary benefits
> cannot compensate for these priceless values.
> The majority of people want an alternative
> lifestyle rather than more (pollution/traffic/
> suburban sprawl/faceless bureaucracy). People
> want to be free from dependence on (Big Business/Big Government/Big Brother). Small is
> beautiful....

Again, no specific quote is needed; this type of political rhetoric is as commonplace as the former.

The element that both unifies and divides the two arguments is the concept of "freedom." Both sides agree that freedom is something to be preserved and hopefully enhanced. Yet each side sees the other as inimical to freedom, for apparently objective reasons. The argument is less over the objective consequences of either the project or its termination, than it is over the evaluation of those consequences. The underlying conflict is profound but the apparent paradox in this case is largely the result of semantic confusion. Actually there are two different kinds of "freedom" at issue; the two historically often have been in conflict and they now have become increasingly incompatible.

The two kinds of freedom are Freedom A, or the freedom of integration, and Freedom B, the freedom of differentiation. Differentiation is equivalent to autonomy, either of a coherent group or of an individual; it is what the Founders referred to as "Liberty." Integration is the formulation and enforcement of the social contract that binds individuals and groups together into larger communities, presumably for protection from the hazards to which the autonomous are vulnerable; it is what the framers of the Constitution referred to as "the General Welfare." The two types of freedom actually have been in contention with each other since the founding of the American colonies. The federalist concept as embodied in the written Constitution of 1789 was an attempt to compromise the two kinds of freedom, but the Civil War was required to resolve the conflict between them in the social constitution.

Speculations on the Transition to Sustainable Energy

That resolution of the conflict has been always somewhat tenuous, and now may be unravelling.

Freedom A is, among other things, the freedom of mass consumption. Freedom A raises the floor of the economic pyramid and brings the poor into the middle class. Freedom A is the central concern of the liberalism of the 1930s, the liberalism of Keynes and his latter-day disciples, the liberalism of large-scale institutions. Rural electrification, the TVA, highway construction, public universities, school integration, radio and later television, commercial aviation, the telephone system, Jones Beach, Disneyland, Hollywood, the NFL, Madison Avenue, Winnebagoes, the AFL-CIO, McDonalds, Sears, Neil Armstrong, and Elvis Presley, to name a few, are all monuments to Freedom A.

Freedom B, on the other hand, is the freedom of individualism, of self-reliance, of conservation; it is also the freedom of elitism, of cultural pluralism, of excellence, of beauty, of uniqueness, of ecological diversity. Freedom B is the principal concern of contemporary libertarianism and environmentalism (social movements of far broader social significance than their formal organizational membership would indicate). Freedom B is a backlash to the entropic, homogenizing impact of the successes of Freedom A. The anti-War (Vietnam) and anti-nuclear movements, CB radio and video tape recorders, Marin County, urban gentrification [21], backyard gardens, bicycles, Laetrile, marijuana, Perrier, the Gossamer Condor, balloon races, "est," diesel Rabbits, communal farms, The CoEvolution Quarterly, Celestial Seasonings tea, the Sierra Club, running, bi-lingualism, "Masterpiece Theatre," E. F. Schumacher, and Robert Redford are contemporary symbols of the yearning for, if not the attainment of, Freedom B.

At the aggregate level, some balance can be struck between Freedom A and Freedom B, at least while the total system is uncongested or until growth is stopped. At the margin, though, the conflict between Freedom A and Freedom B is absolute; that is, an increase in one only can be purchased by a decrease in the other. In the absence of a onsensus about an ultimate and permanent balance between the two freedoms, every marginal change becomes a paramount battle for the survival of "freedom" itself.

As a general rule of ecology, behavioral conflict tends to become translated into territorial conflict. The paradigm conflicts between technocracy and humanism, and between Freedom A and Freedom B, are synthesized in a growing conflict between centralization and decentralization. Thus, centrifugal forces increasingly challenge and strain the centripetal bonds of national and multinational interdependence. Just as dinosaurs were vulnerable to egg-snatching by mammalian rodents, physical and social limits make giant organizations ever more vulnerable to the leverage of small groups. In confronting the mighty, the meek aspire to inherit not the whole earth but only small pieces of it: South Malucca, Eritrea, Croatia, Scotland, Quebec, Kurdistan, or domestically, Maine (Indian lands), Michigan (the Upper Peninsula), Petaluma, or Martha's Vineyard [22]. The way to defuse the destructive potential of irreconcilable paradigm conflict is through division of territory. Historical examples are plentiful: the Biblical exodus, the division of the U.S. into slave and free states, the division of the Church into Eastern and Western branches, the creation of ghettoes, the partitioning of Germany, of Ireland, of Cyprus, of Palestine, and of India, etc. Clearly, territorial division is no panacea for conflict, but in many cases it probably has reduced or deferred some of the destructive consequences of profound conflicts. The impulse to separatism, localism, isolationism, and balkanization seems to be an inevitable feature of the coming transition to sustainable energy.

To Americans, the kind of extreme, polarized conflict I have just described is often difficult to recognize or accept. To a European, the proposition would be quite understandable. Socialists and conservatives, Catholics and Protestants, Hindus and Moslems--to most cultures outside North America such social polarization is conventional. Conflict between polarized social paradigms has been commonplace throughout the world and throughout history. Hegel and Marx were intrigued by it but did not invent or discover it.

Marxism, with its dialectical framework, generally has not made much sense to us in North America. Four hundred years of growth and expansion have permitted continual resolution of such root conflicts (with some exceptions such as slavery) by the mechanisms either of the positive-sum game or of territorial separation. The expanding pie permits you to get at least some of what you want without taking it away from someone else. And if you cannot get along with the dominant social paradigm in New England, you move to Pennsylvania, or Ohio, or Utah. The social geographic map of America becomes not a homogeneous cloth but a crazy-quilt of patches of Amish, Mormons, Norwegians, Irish, Jews, Blacks, Orientals, Chicanos, and so forth. And they get along while there is enough space to avoid mutual interference. So this benign condition, first observed by the historian Frederick Jackson Turner, is a corollary to Kenneth Boulding's more recent notion of the "cowboy economy" [23]. One consequence is that we Americans are unaccustomed to recognizing the possibility of irreconcilable, inescapable social conflict. The end of growing exploitation of space and material resources which must come with the solar age spells an end to such social conviviality and the beginning of an era of conflict.

Paradox

The kind of profound conflict just described by itself is insufficient cause to explain the social schizophrenia which Bateson suspected was required to end our petroleum addiction, and whose early symptoms are now apparent. The "double bind" theory of schizophrenia requires existential paradox, which compounds the stress of intense conflict with important but contradictory demands:

> In the Eastern religion, Zen Buddhism, the goal is to achieve enlightenment. The Zen master attempts to bring about enlightenment in his pupil in various ways. One of the things he does is to hold a stick over the pupil's head and say fiercely, 'If you say this stick is real, I will strike you with it. If you say this stick is not real, I will strike you with it. If you don't say anything, I will strike you with it.' We feel that the schizophrenic finds himself continually in the same situation as the pupil, but he achieves something like disorientation rather than enlightenment....[15]

The current energy crisis already is generating such paradoxical demands. A number of recent incidents illustrate the phenomonon:

- Last year in Washington, D.C., the local utility company announced it was going to have to raise its electricity rates because conservation had been so successful that the company's cash flow had been seriously diminished. This is a classic double bind. Two messages are being delivered to the public. One is that you must conserve energy to save your country and yourself. The other message is that your success in conserving energy will be punished with higher prices.

- Also last year, the Congress passed legislation deregulating the price of natural gas, ostensibly to encourage conservation of this diminishing resource. Within a few months, news reports indicated that deregulation had led to a "glut" of natural gas in the U.S. Gas companies began to recruit new customers aggressively. The federal government, which had been trying to force industry to replace gas with coal in major fuel-burning installations, suddenly reversed policy and demanded that industry increase its use of gas in order to save oil.

- The Austrian government spent about $600 million on construction of that nation's first nuclear power plant. A plebiscite of the Austrian voters--who apparently considered the risks of nuclear power unacceptable--then prohibited operation of the plant.

- In the midst of the gasoline crisis of May 1979, the federal government asked the public to switch from private automobiles to public transportation and simultaneously announced that Amtrak service would be cut 40 percent.

- In the wake of the disaster at Three Mile Island, Metropolitan Edison requested a rate increase to pass the cost of replacement power for the damaged nuclear plant along to its customers. The irate customers complained to the state public utilities commission that the utility's stockholders should pay the full cost of the management's failures in the design and operation of the ruined plant. Met Ed responded that paying for the replacement power for the failed unit would bankrupt the company and drive it out of business, presumably increasing its customers' electricity bills even more drastically.

Further examples of such energy paradoxes could be listed almost indefinitely. Such events are becoming typical of the age of transition to sustainable energy. However, the paradoxical nature of the coming transition is deeper and more complex than what merely anecdotal evidence might suggest.

Following Lovins's notion of hard and soft paradigms, let us identify the constellation of technocracy, Freedom A, and centralization with the "hard" paradigm, and the constellation of humanism, Freedom B, and decentralization with the "soft" paradigm. Were these paradigms truly crisp and complementary, they might provoke political revolution or civil war, but not the kind of social schizophrenia which seems likely to emerge from energy addiction, and which is so mindboggling to the individual person. Actually, these contending paradigms are profoundly paradoxical, both individually and in their mutual relation.

Let us first consider the paradoxes of the "soft" paradigm, and particularly the confusing concept of decentralization. References to "decentralization" are increasingly common in discussions of solar energy policy and program planning. Formal policy studies, research projects, and action programs addressing "decentralization" or "decentralized systems" already are underway [24]. Yet an examination of the literature and activities in this area reveals a high degree of ambiguity in the meaning of "decentralization."

In common usage, the term "decentralization" seems to overlap several other recently coined terms--"soft energy," "appropriate technology," "small scale technology," "renewable energy," and "distributed systems." None of these is precisely defined, but it is clear that their meanings are not completely equivalent. "Small scale" energy systems could be linked in a highly integrated "centralized" network. Conversely, "large scale" energy facilities could be operated in a strictly on-site, "decentralized" mode. Which of these would be viewed as most "appropriate" is unclear. "Small scale" fossil fuel (e.g., coal) energy systems could be more "appropriate" in terms of capital-intensity and thermodynamic end-use matching than would be some "renewable" (e.g., solar) systems. "Small scale" or "decentralized" systems often are presumed more amenable to democratic or local control, but could be owned by centralized banks or utilities. On the other hand, "large scale" cooperatively owned electric utilities are "distributed" and democratic in their ownership. And an integrated network may be viewed either as a "centralized" (e.g., joining a group of electric utilities) or a "decentralized" (e.g., joining a group of farmer, consumer, and labor organizations) structure.

Lovins's definition of "soft energy" included "decentralized" as only one of several necessary conditions, yet "decentralized" or "distributed" often are used as synonyms for "soft." Although "decentralized" sometimes is used to refer strictly to the organizational form of energy technology, a social and political component invariably is attached, sometimes leading to paradoxical combinations. I will cite two examples.

Farmers in northwest Minnesota have been fighting construction of a high-voltage power line across that part of the state for several years [25]. Paradoxical aspects of this situation are numerous. First, the farmers are high technology business people who themselves are major users of electricity. Indeed, rural electrification has been a crucial component of agricultural development. Yet in this case they are fighting electrical power. Second, the utility the farmers are fighting is cooperatively owned by them. Finally, the anti-power-line coalition originally argued that the state government should keep out of this issue and leave the major decisions to local zoning boards. However, when it eventually became evident that the local zoning boards were sympathetic to the power line project and that the state government was hostile to it, the coalition reversed its strategy and argued that the decision about the power line should be made at the state rather than the local level. So how is this to be interpreted: were the farmers in favor of centralization or decentralization?

Another example of the same sort of conflict occurred in Denver, Colorado. For several years there was intense political conflict over the construction of a large scale, capital-intensive water treatment complex called "Foothills." The people who fought the project were generally in favor of limiting the growth of that burgeoning community; they were people who, to a large extent, would identify with the ideas of "soft energy" and "appropriate technology", and hence who presumably would be associated with decentralization. Yet those who opposed Foothills ultimately did so by appealing to the EPA in Washington to block the project. (The strategy ultimately did not succeed in stopping the project but did result in collateral requirements for water conservation.) On the other hand, the proponents of Foothills pointed out that the project had been approved by a referendum of the local voters, and that the project involved local funds and locally owned water and land, except for a very small portion of federal land which would be crossed by an aqueduct from the project. Therefore, the proponents argued, this was a local issue in which Washington should not interfere. So, again, simple interpretation is difficult: which party favored centralization and which favored decentralization?

The "hard" social paradigm is no less fraught with paradox. The essential paradox of the hard paradigm is its intransigent commitment to its own growth--growth in technocracy, growth in mass consumption, growth in centralization of power. Bateson once observed that the French proverb, "The more things change, the more they remain the same," is equally true to its converse form, "The more things remain the same, the more they must change." The status quo that the hard paradigm seeks to preserve is not a static one but a moving, constantly accelerating one. Yet growth inevitably must confront physical and social limits, and must pass a point of diminishing returns and diseconomies of scale. The United States passed that point at least by 1970, when domestic petroleum production peaked and began its inexorable decline. At that point, the transition to sustainable energy had begun [26].

Speculations on the Transition to Sustainable Energy

Even before the hard paradigm began to press its physical limits, the system had begun to exceed the social limits to growth, as these have been brilliantly illuminated by Fred Hirsch, who attributes the self-limitation of growth principally to two phenomena: what he calls positional competition, and commercialization [7].

The fossil energy economy supplies two kinds of economic demands: material and positional. At an early stage of economic growth, material demands (food, shelter, tools, basic education) become largely satisfied. Further growth in demand increasingly shifts toward positional goods--that is, goods (including social relationships) which are "either (1) scarce in some absolute or socially imposed sense or (2) subject to congestion or crowding through more extensive use." [7] Examples of positional goods are an executive job, an ivy-league education, a summer house on a secluded lake, travel to exotic lands, skiing on virgin powder snow, and various symbols of social status. Positional competition is closely associated with the concept of Freedom B stated above.

The paradox of positional competition is the paradox of the "rat race," that is, literally the problem of "keeping up with the Joneses." Members of a growing middle class use their increased income to purchase positional goods previously reserved for a smaller, upper class. Since nearly all members of the middle class do the same, the resulting competition for intrinsically limited goods leads to consequences which individually no one wants. Seeking the "freedom" of rural living, the middle class generates suburban sprawl. Travelling *en masse* to exotic lands, they find Holiday Inns, McDonalds, and other tourists. Attempting to advance their occupational status through education, they create grade inflation and diploma mills.

Economic growth also is limited by a commercialization effect. The best things in life once may have been free but they increasingly become commodities to be purchased. Air pollution makes clean air a commodity to be purchased through the financial investment required by moving to the suburbs and commuting to the city. One result of the social alienation caused by economic growth is that the emotional support once provided by family and friends increasingly must be purchased as a commodity from psychotherapists. As the automobile leads to urban sprawl, the resulting dispersion of social services and amenities make pedestrianship impossible, public transportation uneconomical, and car ownership a necessity. The social limits of growth are intrinsically and painfully paradoxical. There is a universal recognition that sexual intercourse which is freely given is of a higher quality than that which is purchased (at least <u>ceteris paribus</u>, and perhaps even otherwise). In exactly the same sense, the commercialization of public goods into private goods inevitably transforms and reduces their quality, leaving the consumer not only poorer but dissatisfied. The paradox of positional competition is that a mass of individuals seeking Freedom B collectively demand and get only Freedom A, which thereby reduces Freedom B and increases the intensity of the competition.

Moreover, positional competition and commercialization are mutually reinforcing. Positional competition creates a monetary demand for limited social goods which results in their commercialization. But just as prostitution strips sexual intercourse of its romantic value, commercialization removes the positional value of scarce social goods, leaving positional demands not only unsatisfied but intensified. The net result is a largely hidden form of economic inflation which grossly exaggerates the apparent benefits of economic growth. In the hard paradigm, most if not all increase in economic production represents merely increased social and psychological costs of living. In order to remain the same, the system must become ever more demanding, ever more obtrusive, until ultimately it destroys its own reason for being.

Therefore, the conflict between hard and soft paradigms itself poses a paradox, or more precisely, a dilemma. The choice between hard and soft is not one between good and evil, though a few zealots would argue so. For the great majority of moderate-minded individuals, both paradigms contain relatively desirable and undesirable features. Ideally one would like to select the desirable elements of each and concoct an optimum, third alternative. But the components of each paradigm are both internally compatible and mutually reinforcing; at the same time, they are exclusive of the components of the competing paradigm. An economy cannot simultaneously achieve exponential growth and a steady state. Institutional management cannot be increasingly both technocratic and humanistic. Social structure cannot become more centralized and more decentralized at the same time, in the same place. Freedom B only can be increased by a reduction in Freedom A, and vice versa.

An age of paradigm conflict is an age of great anxiety. Reality itself seems to be, and is, in doubt. Individuals who would prefer generally to choose compromise increasingly are compelled to choose between incommensurable absolutes, neither of which promises the security and comfort of the vanishing status quo. The individual's position becomes like that of the Zen pupil. If you say this stick is real, I will strike you with it. If you say this stick is not real, I will strike you with it. If you say nothing, I will strike you with it. The result is profound confusion, and the seeds of madness.

The Madness of Schizophrenia

Social schizophrenia seems to be a necessary part of the cure for energy addiction. The schizophrenic break may be a source of enlightenment for some. However, the potential for social schizophrenia to be manifested as destructive madness is great and ominous. Grave symptoms already can be seen.

Whatever its political cause or justification the revolution in Iran has acquired an increasingly psychotic quality. Arbitrary executions already are epidemic, and social disorder intensifies daily in that troubled nation. The political revolution may be confined to Iran, but its economic, social, and psychological impacts penetrate the U.S. The impact on domestic oil consumption is only the most tangible effect. The U.S. has become the symbol and the scapegoat of Iranian anger. Assassination squads reportedly have been dispatched to the U.S. to execute the deposed Shah, should he ever reside here. Lunatic manifestations of Islamic theocracy if carried to sufficiently threatening extremes could produce a religious and military backlash among the predominantly Judeo-Christian population of America and Europe. Crusades are not unprecedented.

Closer to home, the massacre at Jonestown (whether suicidal or homicidal), though seemingly unconnected to "the energy crisis," may be the most portentous event of the 1970s. Polls indicated that public awareness of the Jonestown massacre (98%) was the highest for any news event since the attack on Pearl Harbor. Jonestown was not merely a bizarre or isolated incident. Jonestown demonstrated the possibility of certain forms of social behavior which American mythology excluded as alien to the national character. Events of violent destruction of others--My Lai, Kent State, political assassinations--however painful, generally have been viewed as occasional excesses of America's aggressive character. But Jonestown symbolized despair and self-destruction: characteristics excruciatingly contrary to America's vision of itself.

Moreover, Jonestown embodied at least to some extent the feudalistic and theocratic tendencies I have ascribed to America's impending future. This is not to say that Jonestown is prelude to a national holocaust. Rather, the substructure of the Jonestown story contains social and psychological elements which are ubiquitous in American society--group pressure, charismatic fascination, anomie, spiritual hunger, incipient parochialism and xenophobia--and which will become increasingly manifest in the coming age of transition. The psychological trauma of Jonestown was felt most acutely in the state of California, whose political and social system elevated Rev. Jim Jones not only to legitimacy, but to luminescence. Jones, the fatal patriarch of Jonestown, was no pariah, no Charles Manson; Jim Jones was a celebrity in California society, embraced and curried by many members of the elite. And as is any member of the "in crowd," Jones was literally above suspicion. The tragedy at Jonestown represents not only the consequences of madness but also the consequences of widespread unwillingness or inability to perceive the evolution of madness.

Another symptom of potential social madness, far less dramatic but potentially just as ominous, is the intransigent faith of the American public in conspiracy theories of the energy crisis. A recent NBC/AP poll (May 1979) indicated that 65 percent of the public believes that fuel shortages are caused by a conspiracy among the major oil companies, despite numerous documentaries, news reports, and official government assurances to the contrary. Cynicism rapidly is deteriorating into paranoia. And paranoia cannot be cured by rational argument. There is no way to prove that a conspiracy does not exist; at best one can argue that the alleged conspiracy is implausible (not easily done given the nature of the oil industry) or that exhaustive investigation fails to reveal facts to prove the conspiracy's existence. But as Carl Sagan has noted in another context, "Absence of evidence is not evidence of absence." [27] To the paranoid, lack of evidence of a conspiracy simply can be interpreted as proof of the conspiracy's scope and effectiveness--"they" got to Cronkite, "they" got to Schlesinger, "they" got to Carter.

There is nothing funny about this. Paranoia and schizophrenia make a deadly amalgam. If the American people lose their trust in the very concept of truth, Jonestown could become a model of the nation's future.

Finally, the most prophetic symptoms of the psychotic potential of the transition to sustainable energy have been evoked by the disaster at Three Mile Island. The psychological significance of this incident warrant an exhaustive study which the State of Pennsylvania recently decided to initiate. Some of the troubling psychological manifestations of this event are:

Speculations on the Transition to Sustainable Energy 211

- many children of Middleton have reported recurring nightmares of the nuclear power plant exploding and killing all the people;

- sales of guns and ammunition increased several-fold in the Middletown area following the accident;

- the community in the general Harrisburg region, and perhaps throughout the state of Pennsylvania, appears to have acquired a "never again" attitude about nuclear power similar to the Jewish sentiment about the Holocaust;

- government and industry nuclear technocrats have developed a standard litany about the TMI accident that "nothing happened" (exactly these words are used most commonly); that, in effect, TMI proved that nuclear power is safe; and that, in any event, there can be "no turning back" (again, exact words) from what is viewed as an irrevocable national commitment to nuclear power—denial, rationalization, and fatalism characteristic of addiction; and

- the general social and political response has focused almost exclusively on the risk of physical cancer resulting from the accident, with little discussion or even recognition of the social "cancer" resulting from the psychological trauma of the event.

To interpret the significance of these symptoms, I return simply to an earlier observation: the unwillingness or inability to perceive the evolution of madness is as dangerous as madness itself.

Conclusions

A few principal conclusions follow from the preceding assessment. First, in the course of the next several decades, the "soft" paradigm is almost certain to become the dominant social paradigm of America. Though the hard and soft paradigms are complementary, there is an underlying asymmetry in their relationship which ultimately favors the soft paradigm. Fragmenting a society is necessarily easier than integrating one. Also, as noted earlier, a common response to social paradigm conflict is territorial separation; but in this case, such a response constitutes a victory for the soft paradigm. And though both paradigms have internal paradoxes, the soft paradigm is more able to deliver what it promises than is the hard paradigm. For example, the hard paradigm convolutes Freedom A and Freedom B in a way which is inevitably self-defeating; the soft paradigm promises Freedom B and works in that direction. Too, there are physical and social limits to the growth the hard paradigm demands; there also may be limits to the soft paradigm's non-growth, but they are at least more subtle [28].

Second, as Lewis Mumford has said, "trend is not destiny." The feudalistic and theocratic tendencies of the transition to sustainable energy need not result in a stifling, inhumane society. These tendencies can be guided toward desirable ends. Many choices are available which may preserve and even enhance democratic and humanistic social conditions. Also, the more feudalistic-theocratic society likely to emerge at the end of the sustainable energy transition need not be a permanent or stagnant state. Certainly in the past such societies have endured for thousands of years with little change. But advanced science and technology are not condemned to disappear during the transition to sustainable energy; the coming "middle age" (literally an age of maturity) need not be a "dark age" of lost knowledge. On the contrary, the emerging "appropriate" technologies for using renewable resources in many cases are highly sophisticated.

The sustainable energy society itself may be transformed, albeit at a slower pace. If scientific and technical knowledge are permitted to continue to expand, the sustainable energy society may well evolve into a higher level of civilization than what preceded it. Such a civilization eventually will renew the attempt to extend its reach beyond the bounds of the earth's surface. Barring a catastrophic collapse of civilization into a truly dark age, human society seems destined eventually to colonize the extraterrestrial environment. Current dreams of space colonies are premature, but may become timely within a century or two. My image of the sustainable energy society then is one of stability, but not necessarily one of stagnation, and certainly not one of hopelessness.

Third, the most critical need for the coming transition period will be to develop a capacity to safely express and manage conflict. A prerequisite of managing conflict is to recognize the fact that it exists. Then a search can be made to find ways to limit the destructiveness of conflict. An inevitable fight still can be a humane and fair one. In the evolution of animal behavior, most successful species have developed ritualized mechanisms of conflict, which reduce the destructive consequences of conflict while preserving its role in natural selection. In many traditional human cultures, similar forms of ritualized conflict evolved over many generations. Unfortunately, in this century the technology of mutual destruction has developed far more rapidly than has the compensating technology for human conflict management.

Finally, however improbable the speculations of this essay may seem, I hope at least that they demonstrate the importance of bringing the behavioral sciences, social sciences, and humanistic disciplines into the assessment of domestic energy policy. So far the technocratic paradigm, based almost exclusively on engineering and market economics, has dominated the analysis and discussion of national energy policies. A principal reason for the speculative nature of this essay is the paucity of investment in behavioral and social research on the kinds of issues discussed here. We desperately need such research, not only to satisfy our curiosity about where our society is headed, and not only to avoid the disastrous pitfalls of the coming age of transformation, but to find the pathways that lead to a sustainable and hopeful future [29].

Notes and References

1. Charles J. Ryan, "Energy and the Structure of Social Systems: A Theory of Social Evolution", paper presented at the Annual Meeting of the American Association for the Advancement of Science, 7 January 1979 (Dept. of Engineering-Economic Systems, Stanford University, Stanford, Calif.); W.F. Cottrell, Energy and Society (McGraw-Hill, New York, 1955).

2. U.S. Central Intelligence Agency, The International Energy Situation: Outlook to 1985, No. CIA ER 77-10240 U (April 1974); Carroll Wilson, Energy: Global Prospects 1985-2000 (McGraw-Hill, New York, 1977).

3. R. Naill, D.L. Meadows & J. Stanley-Miller, "The Transition to Coal," Technology Review (78, 1, Oct.-Nov. 1975); Roger Naill, Managing the Energy Transition (Ballinger, Cambridge, Mass., 1977).

4. G.M. Woodwell, "The Carbon Dioxide Question," Scientific American (Jan. 1978); "Is Energy Use Overheating the World?," U.S. News & World Report (25 July 1977).

5. This view of the energy transition generally is supported by the findings of the energy project at the Harvard Business School reported in Robert Stobaugh & Daniel Yergin, Energy Future (Random House, New York, 1979).

6. Herman Daly, Toward a Steady State Economy (Freeman, San Francisco, 1973).

7. Fred Hirsch, The Social Limits to Growth (Harvard U. Press, Cambridge, Mass., 1976).

8. William Leiss, The Domination of Nature (Brazilier, New York, 1972); Lynn White, Jr., "The Historic Roots of Our Ecologic Crisis," Science (10 March 1967).

9. I use the term "confederation" here to emphasize that both the Soviet Union and the People's Rupublic of China are comprised of historically autonomous nations and provinces which have yielded to centralized control at best grudgingly.

10. Albert Bandura, Principles of Behavior Modification (Holt, Rinehart & Winston, New York, 1969); B. F. Skinner, Beyond Freedom & Dignity (Bantam/Vintage, New York, 1971).

11. Again, see Bandura in previous note. Of course, this is a behaviorist interpretation; the "glue" could be called the "moral order" and "organic solidarity" (e.g., Durkheim). A variety of theories of social cohesion are available. The behaviorist would argue (as do Bandura and Skinner among others) that the behaviorist terminology is the most parsimonious for describing such phenomena. In any event, the behaviorist view is the most useful for the present discussion.

12. Christopher Jencks, et al., Inequality (Basic Books, New York, 1972).

13. The basic case for "the limits to growth" was made in the book of the same name by D. H. Meadows, et al. (Universe Books, New York, 1972) and was presented in technical detail in the subsequent volume by D.L. Meadows, et al., The Dynamics of Growth in a Finite World (Wright-Allen Press, Cambridge, Mass., 1974). Also see Hirsch in note 7 above. Though U.S. Gross National Product grew steadily from 1975 to 1979, the apparent gains have been consumed by inflation. The conventional income accounts tend to obscure the effects of the increased allocation of income to the energy sector, the growth in taxes, and the high inflation in the costs of basic household necessities (food, shelter, clothing, transportation) that has been far more rapid than the rate of increase of the overall Consumer Price Index. When these factors are accounted for, it appears that individual discretionary income in the U.S. almost certainly has declined in the past decade. Some economists

now anticipate an average rate of growth in GNP for the remainder of the century of less than three to four percent, a great reduction from historic post-War levels. Most of this expected "growth" is likely to occur in the primary production sectors (energy, food, materials), reflecting mainly the depletion of nonrenewable resources; the average individual income is likely to be either stationary or declining through the next two decades at least.

14. W. H. Wordsworth, "The World Is Too Much With Us," in Oscar Williams, Immortal Poems of the English Language (Washington Square Press, New York, 1970).

15. Gregory Bateson, Steps to an Ecology of the Mind (Ballantine, New York, 1972).

16. Thomas S. Kuhn, The Structure of Scientific Revolutions (U. of Chicago Press, Chicago, 1970).

17. Peter Berger & Thomas Luckman, The Social Construction of Reality (Anchor, Garden City, N.Y., 1967).

18. Amory Lovins, Soft Energy Paths (Ballinger, Cambridge, Mass., 1977).

19. Daniel Bell, The Coming of Post-Industrial Society (Basic Books, New York, 1973).

20. Daniel Bell, "The Post-Industrial Society," in H. Kahn, The Future of the Corporation (Mason & Lipscomb, New York, 1974).

21. "Urban gentrification" refers to the rehabilitation and rehabitation of central city housing by middle class individuals (often singles or childless couples) which has become the most significant trend in U.S. urban development. Though offering great hope for urban revival, gentrification has created a serious problem of displacement of the poor from urban neighborhoods.

22. Leopold Kohr, The Breakdown of Nations (Dutton, New York, 1978); Ernest Callenbach, Ecotopia (Banyan Tree Books, Berkeley, Calif., 1975); E. F. Schumacher, Small Is Beautiful (Harper & Row, New York, 1973).

23. Kenneth Boulding, "The Economics of the Coming Spaceship Earth," in G. de Bell, ed., The Environmental Handbook (Ballantine, New York, 1970).

24. Jim Quinn & Jim Ohi, <u>A Compendium of Decentralized Studies, Programs, and Projects</u>, Review Draft (Solar Energy Research Institute, Golden, Colo., 14 May 1979); an important example of this type of research is Mark Levine, et al., <u>Distributed Energy Systems in California's Future</u>, U.S. Dept. of Energy, Office of Technology Impacts, No. HCP/P7405-01 (Washington, D.C., March 1978).

25. Luther Gerlach, "Energy Wars and Social Change," in S. Abbott, J. Van Willigen & G. Neville, <u>Proceedings XIII of the Southern Anthropological Society</u> (U. of Georgia Press, Athens, Ga., Fall 1978).

26. I am indebted to August Giebelhaus for pointing out that this paradox is not new; it was described by Herbert Croly as "Jeffersonian ideals achieved through Hamiltonian means." This dilemma of liberal politics was institutionalized in FDR's New Deal. The significance of the paradox has been augmented, though, by the increasing confrontation with physical and social limits to growth.

27. Carl Sagan, <u>The Dragons of Eden</u> (Ballantine, New York, 1978).

28. The potential limits to "non-growth" are suggested in Ursula LeGuin, <u>The Dispossessed</u> (Avon, New York, 1974).

29. I am grateful to Barbara Farhar, Sylvia Forman, August Giebelhaus, Dennis Horgan, and Michael Yokell for their helpful comments on the draft of this article. The views expressed here are entirely the author's own, and do not necessarily reflect the opinions of any other individual or organization.

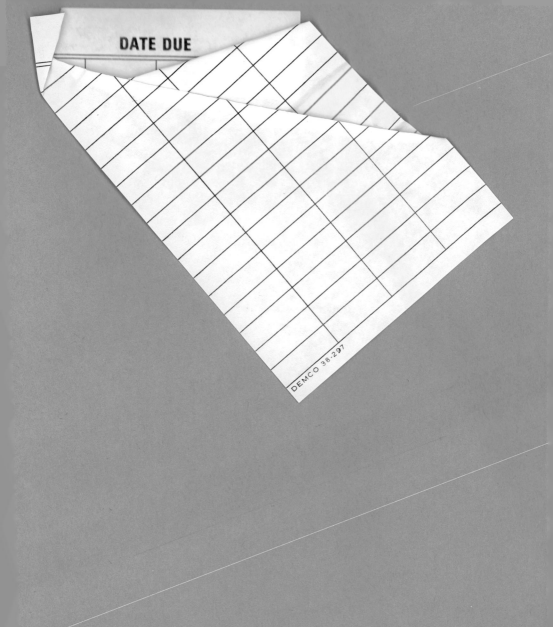